# THE ELEPHANT AND THE BEE

Jess de Boer

# THE ELEPHANT AND THE BEE

Jess de Boer

JACARANDA

First published in this edition in Great Britain 2016 by
Jacaranda Books Art Music Ltd
Unit 304 Metal Box Factory
30 Great Guildford Street
London SE1 0HS
www.jacarandabooksartmusic.co.uk

A CIP catalogue record for this book is available from the British Library.

ISBN: 978 1 909762 24 4
eISBN: 978 1 909762 25 1

Book design by Branding by Garden, London, UK

Printed and bound in Slovenia by Imago Publishing Limited

# PREFACE

*"Don't ask what the world needs. Ask what makes you come alive and go do it, for that is what the world needs, people who have come alive."*
Howard Thurman

At five years old I encountered my first bee. To be more precise, I stepped on it with a bare foot and as the poison spread throughout the limb I thought I would probably die. It hurt like hell and within ten minutes my ankle had swollen to twice its normal size. But it wasn't until several hours later, having suckled the teat of my parent's sympathy dry, that I discovered that the offending bee had fared far worse. For in a final act of defiance her tiny, barbed sting had stabbed through my skin and caught and as she struggled it had yanked out part of her abdomen leaving her to die alone in a dusty section of clover scattered lawn.

I recall feeling dismayed at this news, mostly because my highly convenient victim status no longer stuck, but there was something else too, a sense of awe at the stripy little beast whose morning of purposeful flower bopping had been, quite literally, trampled.

I still get that same feeling of wonderment now, some twenty-five years on as I walk towards a hive (always from behind as honeybees get flustered if you block their entrance path) admiring the single-minded intensity of their comings and goings and occasionally I think back to that first encounter and wistfully appreciate the cloak of ignorance that surrounded those early days of childhood.

But perhaps above most things that I have learnt from the bees, it is their unquestionable sense of busyness that I appreciate most. Because that's where we differ, the bees and I, in that after all this time - and as I approach the dizzy precipice of adulthood, I still don't really know what I am doing – what I want to do with my life and for most people in this frenzied modern world of ours, that is grounds for genuine concern.

Before I discovered beekeeping though, things were far worse as I, like many of my peers, grappled with the potent combination of hormones and homework followed by curriculum vitae's and career advances and just about everything else that adds to the noisy crescendo of 21st century life. And in such a state of bewilderment the very first words of this book were written - in a desperate attempt to find some clarity,

And five years on, here we are.


This book has taken me on a gloriously wild ride and with conviction slowly ushering out the lingering shadows of personal insecurities, let me boldly state the opinion that a single human life consists of no more than an assortment of events and experiences that flow together in one seamless episode and which often requires a moment of hindsight to muse at the glorious haphazardness of it all. Perhaps someone who is less afraid of sounding like one of those self-help pundit's might ascribe the term 'Journey' and on the surface that is what this book is about; a collection of hard-won tales that describe amongst other things what it is to be a young person in the world of today faced with the future of tomorrow.

But beneath the erratic descriptions of this quest for fulfillment it is also my intention to provide some insight into how we might better guide other young people forward by helping them find their own voices too. For it is only as a collective force, much like a swarm of wild bees searching for a new home, that we as a generation might also hope to set out into this big, wide world we one day stand to inherit and have a go at starting afresh.

Confiscating the atmosphere of doom and gloom that clutters so much of today's conversations, this is the story of a bumpy road to self-mastery, one that ultimately lead me into the glorious world of African beekeeping. I came to appreciate that it is only through the course of reconnection with the natural world that we humans might find some stillness.

Today I must stumble

# CHAPTER ONE

I remember the last day of high school like it was yesterday, barreling out of the school gates with insides that fizzed like the top of a freshly poured Coke. As I whooped and cheered with my peers I remember thinking that finally, finally, my life was about to begin. Turning round for a last look at the familiar buildings I waved a fond farewell to the toothless watchman who stood guard at the front gate and blew a loud obnoxious raspberry at anyone who cared to notice.

I had decided to continue my studies and after a series of lengthy applications, was accepted into the University of Cape Town for a broad undergraduate degree in Environmental Science. The University term began in March the following year - that left me with four whole months of freedom in which to do something fabulous.

I recall staring blankly at the spinning globe Dad kept in his office, trying to figure out where I would go and what I would do there. Those exotic countries I had learned about in geography suddenly became possible destinations and I was struck numb by the size of the world we live in today. Mum and Dad were eager for me to get out of Africa, to see the world beyond these third world borders, and I began to research my options. A few days before the last exam, one of my closest friends handed me a small paper brochure from a company she had discovered called Contiki Adventures. It turned out that they specialized in "European Budget camping tours for 18 – 35 year olds ... 'visit 14 countries in 21 days'" and the next trip departed in just over ten days.

Excellent! These three words sold it for me:

- Budget (Mum and Dad would pay for everything),
- Camping (my favorite thing in the whole world),
- and Adventure (yes, please).

*It sounded too good to be true.*

After a quick family discussion that evening and a phone call to the Contiki office in Europe, my place was secured. As the departure date loomed ever closer, I would take out my airline ticket from its sleek blue case and daydream. *Ahh, the rivers we would cross, the great mountain ranges we would climb and the animals we would see!* I couldn't wait.

I was not entirely naïve. I knew that Europe was not wild and untamed like the parts of Kenya and Tanzania we had frequented on safari, but in my dreams there were black bears and timber wolves; they had the Alps and I had scoffed enough Swiss chocolates to know there were some pretty fabulous looking lakes and rivers to be explored.

**A glistening world of possibility and adventure awaited.**

As we didn't have the Internet at the time, I was unable to research deeper into this heaven-sent bus tour, but such was my adolescent innocence that even if we had, I probably wouldn't have bothered. The idea of spending minutes, let alone hours in front of a beeping plastic box was beyond the imaginable. Instead, in the days prior to my departure I carefully started to acquire all the necessary items required to integrate with my fellow European campers - that included:

an assortment of pocketknives,

a dismantle-able catapult,

a butterfly net,

a potato gun,

a camouflage stick,

a fishing rod

and several khaki shirts that I wouldn't be seen dead in at home, but that I thought would impress my European fellow campers.

*I was coming from Africa, after all.*


The evening of my departure I was taken to the airport by my entire family - brother, parents, grandparents and a crap aunt called Nance that no one really liked and who smelt of molasses. I was fussed over and embraced to within an inch of my life and in case I got hungry, Mum even packed me a boiled sweet potato (my favorite) to nibble on during the flight. With much waving and running down the up-only escalators at Jomo Kenyatta International Airport for the final goodbye, suddenly I found myself alone. I recall feeling rather small and although I didn't qualify for the children's coloring book on British Airways, the sweet-faced female flight attendant treated me like a princess. That was until the aforementioned sweet potato rolled out of my bag and was run over by the food trolley about an hour into the flight. Kind of funny if you knew what it was, but to the uniformed cabin crew it must have looked like the end result of an awful Chicken Korma experience.

The next few days after I landed in London were hazy; distant cousins picked me up, I ate something called Spotted Dick and experienced Marks and Spencer's for the first time. Never before had I seen so many white people, fat people, and after discovering the pick-and-mix sweet section in a shop called Woolworths, I seriously doubted whether I would ever return home.

Then before I knew it I was being dropped off by the translucent cousin at the entrance to a large, skulking hotel on the outskirts of London from where my so-called camping trip departed. I hadn't slept for days and was so full of nervous energy that when combined with the smooth roads and air-con in my uncle's shiny automatic I felt sick, really sick. Somewhere along the M25 after a bizarre breakfast of pop tarts and pink Nesquik I found the absence of wind-down windows too much and puked on the sat nav.

Arriving at the departure point will haunt me forever. I stepped out of the car and grabbed my backpack with trembling fingers, poking my uncle in the eye with my fishing rod as I did so. Hitching up my cargo shorts I took a final deep breath and turned around my gaze settling on a mass gathering of young people who had accumulated outside the reception area of the towering hotel. Excited chatter filled the air but I immediately noticed something was wrong. Where were the hiking boots? The khaki? Why was that girl with a wheelie-suitcase wearing a short black dress, and what was that they were drinking?

Surely not beer?

But there was no turning back.

Smiling awkwardly I slipped out of my multi-pocketed safari jacket and edged my way around the group to what could only be our bus boldly emblazoned with the purple CONTIKI logo along the side. Hope flickered like a candle in a monsoon as I bagged the front seat next to the driver - brilliant, I would spot the animals first - and, whistling nervously, I placed my ham sandwich and accompanying paper napkin in the seat pocket in front of me.

It was when I was wrestling with an unruly butterfly net that I noticed faint scuffling noises coming from the back of the bus. Believing I was alone I went to investigate, pocketknife at the ready. At home buses like these were often targets for troops of baboons who invaded in the search of food and Japanese people and, although I knew this wouldn't be the case here, I did half expect some form of creature life to be the cause of these strange sounds. I really don't want to spend too much time describing the results of my first explorative mission on that horrid bus in this foreign land, but suffice it to say that school biology lessons were poor preparation for my first experience of actual, real, live human copulation.

*It was 9:30am on the first day of the longest, cruelest three weeks of my life.* The Eiffel Tower, the Louvre, the Vatican, and several cathedrals were amongst some of the supposed 'highlights' of the trip. But looking back at blurred photographs I remember very little. Most of these famous sights were viewed with red eyes and a churning gut. Here's a little review of the daily routine…

**8:00am:** Lurch around bus competing for the honor of that day's worst hangover while waiting to leave the campsite. Enter Highway.

**12:00pm:** Stop at a petrol station for lunch. Throw up in flowerbed.

**12:30pm:** Re-enter highway, head for a border.

**5:00pm:** Arrive at new campsite. Put up tents, eat supper, enter bar. (Sometimes we got taken to an evening show that was included in the price of our trip. Such side excursions included a sex show in Amsterdam and a rip-off version of the Moulin Rouge in Paris.)

**3:00am - 5:00am:** Go to sleep.

**8:00am:** Wake up, force breakfast down and board bus for more lurching.

**12:00pm:** Drive to a central location and get off bus. Wander around. Get lunch.

**4:00pm:** Board bus, return to campsite. Begin drinking

**Repeat.**

To be fair, the three weeks passed quickly. Apparently we visited *14 countries in 21 days* and everyone else loved it. I was just different and at the time wished with my whole heart that I wasn't. When one Australian puked down the side of the Tivoli fountain, I wanted to cry. Not because it desecrated a sacred monument, but because I too wished I could puke! I still hadn't figured out how.

Back home I hadn't really drunk alcohol, but it was all that seemed to happen on that trip. I hated the smell, the taste and the expense of it all, but so desperate was I to fit in and play the drunken clown that I continued to fork out fistfuls of my parents' money to local barmen in return for yet another bottle of luminous green liquid. Unfortunately, my efforts were never quite persuasive enough and I remained the token bus weirdo for the entire trip. Weird doesn't go down too well with teenagers.

The only time I look back on with a smile was when we had stopped for a novelty passport stamp and lunch in the tiny country of Lichtenstein. I had bought a piece of bread and a tin of sardines from a small shop for the equivalent of a term's school fees back home and sat alone on a bench bordering a park. I was doing my best to wipe the resulting oil spill off my shirt when I glanced up and caught the strange sight of a Maasai warrior standing tall and still in crowds of weekend shoppers.

Seriously doubting my sanity for the eighth time that day, I made my way over and introduced myself. Clad in the traditional red *shukka* - shoes made from car tires - multiple strands of beads and red ochred hair was a proud *Moran* or young warrior, whose name was Ben. We struck up a conversation and through a garbled mixture of Swahili and a few basic words of Maasai I learned that Ben had arrived in Europe to seek his fortune and was doing well selling various strings of beads and posing for photographs with tourists. Most of the money he did earn was sent back home to his family who lived in the village of Narok, a bustling township that borders the edge of the famous Maasai Mara Game Park and although it can't have been easy living so far from home, his family and the equator's warmth, he was happy. It gave me a surprising amount of strength to see this man, standing alone and proud, making the most out of being different.

I can't recall when I gave up and stopped drinking on that tour. I had tried so hard to fit in, but ultimately I was an outsider. I didn't have cool clothes, I couldn't handle more than two beers without feeling queasy and the smoky depths of the campsite bars made my eyes water and my throat sore. I hated sitting on the bus for ten hours without access to fresh air and I was not interested in having sex with boys. At first this attracted the attention of the bus community and people would ask to sit next to me to hear my heavily exaggerated stories from home, but when I started going to bed alone and sober on day six, I instantly reverted back into being a 'loser'. Once you've stumbled down that road, there appears to be no turning back.

Ten days into the infamous bus tour I came across a book at one of the plastic campsites we habituated for a night. The site itself was another nasty, crowded piece of land this time bordering the Amsterdam sewerage ponds and the book was the kind read by those who stayed there often. It was a sex novel and, abandoned once again by the group, I read every page. It kind of inspired me and I came up with my own:

### The best sex scene, like, ever:

*"His grubby hand tapped the underage, fat girl on her shoulder. She turned slowly. His breath caught in his throat; she was by far the ugliest girl on the bus, and the drunkest too - he was in.*

*She swayed on her feet, and burped under her breath. He took a deep breath and steadied his blurry vision. His resolve crumbled. She was fucking awful. But then he glanced up: the campsite bar was heaving and three people deep. No chance of getting another beer and the ratio of males to females was not in his favor. The dance floor was full of sweating, gyrating German camper-men who had left their wives alone back in the caravans watching television. His mono brow unclenched. She was the one.*

*He raised his greasy eyebrow in what he hoped was a 'come hither look' and uttered the single word that had worked every night since Paris: "Heeeyyy". She was taken aback by his sudden keenness, made stranger by the memory of him sneering at her in front of the leaning tower of wotsit, or was it the Eiffel tower?*

*Whatever - she was in.*

*Fluttering her heavily made up lashes, and blushing slightly beneath a forest of acne, she leaned in. Clutching her yellow Alco pop tightly (3 for 1 at the bar) she shyly whispered, "Hiiiii", and shuffled closer, brushing up against his man boobs. Wrapping his arm around her broad shoulders he led her away from the loud thumping Euro pop, out into the clammy night air. Wafts of beer, barbeque and the sewerage ponds met them under a strip light sky.*

*He made a grab for her tit. She, for his neck. Impact was made and as they mauled each other's upper bodies (they were only eighteen) they stumbled to a darker part of the reception area. Urged on by hormones and desperation they ravaged each other for several seconds, before falling through the dividing hedge that bordered the toilet block. He fell first, she next - all 70 kilos of young flesh landing on his abdomen. "OOOFFFF," he gasped and then looked up, struggling to suppress the sudden urge to chunder. Out the corner of his wobbly vision he caught a movement; a disapproving family was hurriedly washing up their baked bean dinner in one of the sinks. Glancing down, he stared at the girl, she was lying face down between his splayed legs in the basin run-off. Groaning slightly, she dry retched. After several seconds she lifted her gaze to meet his. In the distance an advert for car insurance played from someone's camper home television. A lonely baked bean slid down her cheek. Breaking eye contact he quickly looked around, the washing-up family had gone and they were alone at last. An awkward silence fell. She began to extract herself from the drain, but catching her eye as she hovered on her knees he leaned forward and whispered, "You know, while you're down there..."*

\*\*\*\*

Then we entered Switzerland – oh, the Alps! After several days spent navigating tourist crowds in large city centers, we started to climb into those blessed mountains. At last, green space, silence and as far as the eye could see chains of shimmering peaks. I went to bed extra early that night, determined to make the most of the next day which had been designated as a campsite rest day; **we were free to do as we pleased.**

I had managed to persuade another girl to join me and waking early, we borrowed two bicycles from the campsite owners and off we pedaled. It was the best day of the entire three weeks and having followed a map, we wound up on a rolling cycle path that led us through a huge green forest complete with a carpet of wild flowers. Halfway along the track we came across a group of locals picking small black berries from a number of scrubby bushes and several meters later we pulled up at a fresh clump and sampled the free delicacy. The berries, I later discovered, were called *Myrtilles* and they were delicious.

Having eaten our fill, our blood now surging with antioxidants, we continued on down the track until it abruptly ended on the edge of one of those aforementioned famous electric blue lakes of the chocolate-box variety. Surrounded by clumps of nodding daffodils, the color of sunshine, we quickly changed into our swimming costumes and hurled ourselves into the icy cold water, disrupting the peace and solitude of the scene with shrieks of joy. We then stretched out in the soft, green grass, bodies tingling, and wolfed down pieces of bread and dribbly cheese that had been filched from the breakfast table that morning.

Returning to the campsite later that afternoon I felt refreshed and invigorated, exercised and alive. Everyone else had stayed put around the bus all day nursing their hangovers and when we returned, we were instantly surrounded by the very same people who had tried so hard to ignore me over the past few days, wanting to hear where we had gone and what we had seen.

For over thirty minutes (until the bar reopened) I bathed in the attention of having a story to share with people who had none, thus serving as my first lesson in the contagiousness of positive energy.

Thankfully for everyone involved, the tour ended shortly thereafter and still in one piece I returned home, none the richer, but infinitely wiser to the ways of the world. Moving to Cape Town to begin my degree studies suddenly did not feel quite as daunting as it once had.

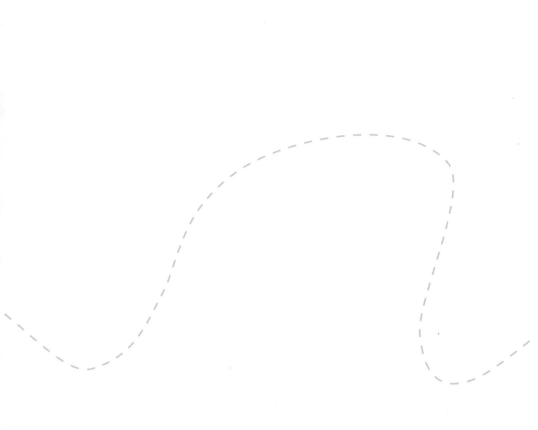

# Today I must begin

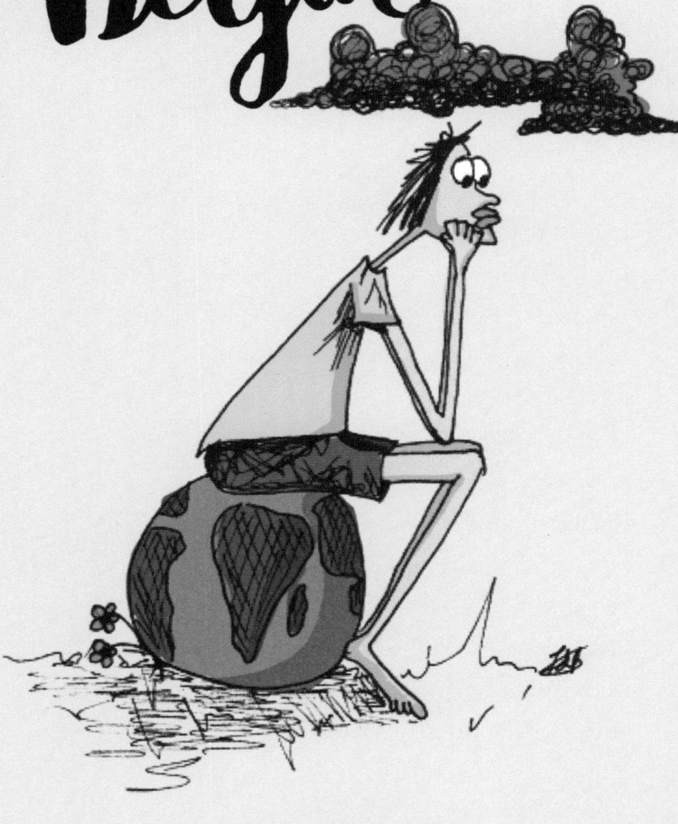

# Chapter Two

Whistling on through time I found myself sat at my favourite table right next to the window, the faintest of breezes occasionally wafting through bringing aromatic hints of the icy cool Atlantic and fynbos the vegetation that covered the slopes of Table Mountain. My classroom was perched directly beneath the soaring summit of Devil's Peak and if I turned my head to the left – at an angle that from the front I hoped portrayed a look of deep contemplation - I could just about make out the glittering waterfall that danced its way off the yellow rocks down into a gully of waving ferns.

I forced my attention back to my lecturer Mrs. Holloway who was once again prophesizing imminent environmental collapse from the front of the dimly lit room.

I was three months into my fourth and final year at the dazzling University of Cape Town and as my earlier undergraduate years had been squandered mostly in the pursuit of mountains, waves and boys I only now realized that I lacked a clear trajectory out into the real world. Alarmed at my apparent lack of application I chose to settle on a very austere sounding, **'Double Major in Disaster Risk Management & Food Security'** and from the very first moment of the first class it had been a rather bleak and dreary ride. That first lecture had provided an overview of the year ahead and I recalled the initial impulse to cover up my ears and hotfoot it to the Fine Arts building, where I was sure to make my fortune in papiér maché; every cell in my being screaming the words, "get out, GET OUT NOW."

Unfortunately on that particular morning I had chosen to sit right in the center of the lecture hall and was closed in on either side by serious looking students scribbling notes. There was simply no escape and re-focusing on the lecturer, a small man with a monotonous voice I sighed deeply, picked up my pen and scribbled down the following list:

Imminent starvation,

water wars,

climate change,

drought,

flooding,

famine,

disease,

swarms of locusts,

rising sea levels,

mass starvation and

collapsing food webs.

Looking over at my neighbor's notes I had then added:

Overpopulation,

the obesity epidemic and

universal extinction

… Apparently things were not so peachy on planet earth.

Suddenly the classroom turned black as Mrs. Holloway turned off the lights and the familiar vortex of doom that had begun to suck away at my soul dissipated as the projector whirred to life. Glancing down at the neon hands on my watch I noticed we had ten minutes left of the lesson and guiltily I straightened my slouch and gripped my pen a little tighter, determined to commit at least one piece of information to memory. After a few seconds of fumbling an image of a field covered in hundreds of blossoming apple trees appeared on the wall complete with rolling green hills that faded off into a pale mauve sky. More determined than ever to spend the remaining few minutes focused on such a pleasant scene I studied the image a little closer and noticed a number of people precariously perched up on ladders elbow-deep in the foliage, their heads covered in those funny conical hats that placed the scene in Asia.

A small sub-heading on-screen read:

**Case Study 7 – How important is a bee? by Robert Krulwich**

Wicked! - I rather liked bees, those wholesome, hardworking creatures that caused Winnie the Pooh so much merriment. But before my imagination gamboled down a road festooned in peanut butter and honey sandwiches the picture on the wall changed to a close–up image of two of the men up their ladders, their loose clothing bleached by the sun and their expressions deeply focused on the job at hand. Both were well up in the flowering canopy of the tree, one hand firmly gripped to the top of the ladder whilst the other grasped some sort of short stick.  With another click the picture zoomed in closer to the stick in question – a paintbrush no less, surrounded by an out of focus blur of rosy, white petals. Rather wishing I had concentrated a little more at the beginning of the lecture I studied this puzzling scene for a few more seconds as Mrs. Holloway began to read out aloud in her soft, mysterious voice:

This story begins in central China, in an apple-growing region called Maoxian County, near the city of Chengdu. In the mid-1990s, the bees that regularly showed up there every spring suddenly didn't. Apple farmers, obviously, need bees. Bees dust their way through blossoms, moving from flower to flower, pollinating, which helps produce apples in September.
What happened to the bees?

A number of explanations were put forward by scientists to suggest why the bees had disappeared but whatever the reason, the bees went missing and the farmers had to do something, and do it quickly. So they decided to replace bees with humans and they began to pollinate by hand.

In 1997, Maoxian apple growers, using brushes made from chopsticks and chicken feathers (and sometimes cigarette filters), went from blossom to blossom — just as bees do, to spread pollen. People worked full shifts, moving up the hillsides as each orchard hit blossom-time...

Furiously scribbling down notes I wasn't able to comprehend the sinister nature of this story until the lights had been switched back on and the sound of zipping pencil cases filled the air. Bewildered I looked back down at my notes and felt something shift deep inside; an all-consuming bolt of panic and disorientation that rose from the pit of my stomach like an ocean wave on first contact with a shallow reef. This was something new, something recent – not like oil spills or climate change (boring) but this was about little bees, innocent creatures upon which so much life depends. *Wait. Stop. Where did they go? What's happening to the bees? How can we let this happen?*

Staring wildly around the classroom as the last of my jostling, backslapping peers shuffled their way out the door into the sunshine outside, I sat back numbed, mouth agape and skin suddenly cold, playing back the image of the men up in the lifeless treetops poking their paintbrushes roughly into the blossom over and over again until a slight puff of fresh air drifted in through the window and ruffled the pages on my desk.

*There really is no hope.*

And that right there is how I left it, for what else could a young kid like me possibly do about the bizarre incidence of the vanishing bees?

* * *

Student life whistled past. With the academic guidance of my university professors and volumes of course material showcasing the extreme scenarios of human shortsightedness my eyes were forced ever wider to the gloomy prospect that represented the next seventy years of human life on earth. Beyond that I simply didn't have the capacity or the will to imagine. Africa, the North Pole, Europe, China and the Americas - everywhere we looked it appeared that entire eco-systems hung in tattered shreds and with each essay I submitted or case study I tickled the surface of, I began to feel increasingly like a young hobbit who had just poked its head out of its cozy burrow, blinked twice and taken in the sight of Mordor rising out the swirling red mists ahead.

But I must be careful here not to lead you to the false conclusion that I spent every waking hour gazing out into the middle distance, all-consumed by the thought of a world without trees or rhinos or bees for that matter, because I didn't. I was your typical student, albeit of the privileged variety, who ate beans on toast three times a day and handed her essays in seconds before the deadline, content to nibble away at the pre-conceived boundaries of the self and the world and quick to shrug off messy concepts such as 'the future' in favor of particular members of the opposite sex, budget groceries and adventure.

It was to this last particular aspect of life that I dedicated most of my time given that this particular part of the world was, and still is, well endowed with opportunities to explore. During the week before, after and on occasion in place of my lectures, I would mount my trusty steed - a bright butter-coloured scooter dubbed The Yellow Peril and putt-putt my way all over the Cape Peninsula. Parking her up at the gates of the magnificent Kirstenbosch Botanical Gardens I would then clamber my way through the scrub and an hour later find myself lost amongst the herb-scented clouds atop Table Mountain.

Then there was the seaside option; a never-ending smorgasbord of sandy bays and enclaves stroked by the icy-cold Indian Ocean on one side and the Pacific on the other. Diving, mussel collecting, swimming, surfing or if the wind was right I'd pump up a bright green inflatable kite bought in a fit of questionable intention which would drag me, wind assisted, often screaming, across the perfectly oiled rows of sun worshippers laid out on the beach like driftwood and out into the surf.

But perhaps my favorite of all were the mountains... Ohhh those mountains! All within a few hours drive from the smoking mother city and the preferred destination for any amount of free time longer than twelve hours. It was up there, amongst those soaring, silent peaks, alongside friends and the occasional pet pooch where I was able to sit and reflect on the microscopic world laid out beneath my feet, silently absorbing the breathtaking beauty of this epic continent. Hiking, peak bagging, kloofing, fishing, bouldering and once or twice when a freak cold snap hit the higher peaks in the Ceres valley, even skiing. Up there I would shed my city-stained skin and feel most at peace, all those noisy thoughts and questions concerning the upcoming decades temporarily silenced as together we breathed in the perfection of the moment. Returning home often late on a Sunday evening, bodies spent, covered in scratches, grassy burrs and smelling of wood smoke I would clear a path through the empty beer bottles left behind by my house mates and send a silent message of appreciation to whoever it was that was listening, grateful for the life I had been granted and more determined that ever to live it large - to max it out – to go big or go home. For I figured that as long as there existed wild space such as that I had experienced, then there was still something left to fight for.

*

Thanks to that final year and dissuaded by just how hard it all looked I came to the conclusion that saving the world wasn't a realistic career choice. Confronted by this certainty I had to put it to good faith that something else would pop up instead; something splendid and worthy of my upbringing, something that would take me by the hand and lead me skipping down a path lined with daffodils, straight into a flaming orange sunset … no sweat man, cool,

yeah bro,

chill.

# Today I must earn a living

# CHAPTER THREE

At the time I received my degree certificate, embossed with the heading 'graduate' in slanted gold lettering I was little more than a statistic. A tiny blip on the radar of a society I still felt grossly ill prepared for and just one member of a 300-million strong army of global employment-seekers lined up before the start of a race like a vast pack of whining, snarling greyhounds straining at the gates in breathless anticipation.

For the first time ever I felt myself infected by intolerable levels of excitement at the thought of what lay ahead. But having never had to run too fast nor fight particularly hard for anything in my short, swaddled life I was apprehensive of my ability to keep up with the others, a doubt further compounded by the fact that other than a rather grandiose notion to 'save' the world I had no direction. Fortunately for me, the ugliness of this situation was postponed thanks to the resurfacing of an old African connection that saw me wobble into my first job just three days into this newfound freedom. The location for this initial foray into the working world kicked off deep in the French mountain-scape with a feather duster as my weapon of choice. Excited by the novelty of this adventure and boosted by the fact that I would be earning a regular income (in euros) I decided to relax my grip on the words 'meaningful existence' and make the most of what I hoped to discover along the way.

*

# The tale of an Alpine cook

I arrived early one December morning at Geneva airport. Striding into the arrivals hall in my favorite 'look at me, I'm from Africa' Bob Marley trousers I was immediately confronted by hordes of politely jostling people dressed in winter black. From their gloves to their boots, hats to briefcases this army of morbids stood around a sparkling arrivals hall in well-brought-up silence, awaiting the delivery of their baggage. A scene so entirely different to the third world chaos I had left; where *miraa*-chewing, AK47-toting security guards milled about disheveled tour groups and herds of rotund, shiny officials nipped at the heels of those who had lost control over their wobbly-wheeled luggage trollies. It took several minutes of continuous blinking to comprehend that everything was in fact okay.

Taking a hint from the people around me, I too politely jostled my bags off the conveyor belt glancing outside through the polished glass windows as I did so. The swollen black clouds I had flown in amongst had stealthily joined forces since we had landed, turning the inside of the airport as dark as a warthog's den.

It was most exciting.

I made it through Customs without being searched for dope and was met at the door by my new co-worker, James, who had arrived the day before from England. After a night of pastis and cheese with the boss, James had been discharged in a large, expensive van to pick me up. He looked fairly rough and announced he felt it too, but as the season progressed I learned that this too was perfectly normal.

Our trip back up the hill towards the resort was like being in a fairytale. Snowflakes drifted down on to the van's sparkling bonnet, orchestrated to the sound of tires swishing through half a foot of icy sludge. Pressing my greasy face up against the window I stared at the cows in the fields --five times the size of Kenyan ones-- all huddled together munching hay as steam drifted off their glistening backsides. I could only fathom the reactions of the people back home at the sight of such enormous beasts - like palomino buffalos, only bigger.

As we climbed further up the hill, the snowflakes thickened and we entered a narrow twisting section of road through a pine forest. It was James's first season in the snow too- he had quit his stressful job as a financier in London and he eyed the whitening road with increasing agitation as the wheels slipped and slid, struggling to gain traction. After ten minutes or so we came to a slushy halt. James looked nervous. I blinked and helpfully watched the windscreen wipers.

The road was empty and the sun had begun to sink although it was only 4pm. James got out and rummaged in the back for something he called 'snow chains' as I twiddled with the buttons on the dashboard. James eventually found the chains and after fifteen minutes of finger-numbing guesswork that resulted in nothing more than a blue nail, he asked for help. We eventually worked out that one of the chain clasps had broken and the prospect of a night of howling wolves and frostbite forced us to double our efforts. After fifteen minutes or so I had an epiphany and after rummaging around in my suitcase, I instructed James to step aside and set about tying the snow chains together with hair bobbles.

Thus, the very first life lesson I learned was this one:

**Always carry string.**

\* \* \*

And so the season kicked off. I slowly got the hang of Alpine skiing and found myself involved in chalet politics when the chef shagged the barwoman followed in rapid succession by the nanny. I learned to appreciate *fondue, merlot, jus and girolles, tarte aux framboise* and enormous logs of creamy goats *buche* and discovered the highest level of euphoria after spanking a weeks' worth of wages on a Chocolate Chaud Royal at Chez Martine's halfway down a blue run.

With the wages I squirrelled away each week I was able to purchase a pair of randonée bindings that enabled me to walk uphill in the snow. Growing tired of the glitzy, diamanté crowds that cluttered most of the main slopes I began to venture off with my new equipment further and further into the endless stretches of Narnia like-forests, swishing across frozen rivers, into prehistoric ice caves and past several herds of chamois goats that clattered up slabs of vertical rock like mammalian geckos. Once I even caught a fleeting glimpse of a young Bucatan - a rare goat-like creature with huge curling horns that stood proudly atop a rocky ledge like the bovine version of Alpha Blondy. Relishing the peace and quiet of these sojourns I soon learnt to cherish the overwhelming sense of contentment that filled my soul right to the brim; self-assured of the path I had chosen and at peace with the world at large, devoid of street kids, belching factories and the congested roads of my previous life.

But all great things have a habit of coming to an end and suddenly it was April. The snowy peaks swiftly turned brown and after half a year spent living in a wintery world of monochrome interspersed by the occasional dash of neon ski jacket there was a tinge of color, warmth and daffodils. As the sun turned the Alpine pastures green, I witnessed my little village in an entirely new light and my plan to return back home to Africa were put on hold as I frantically looked for work that would keep me *in situ*.

In the months leading up to spring the chef at my chalet had become increasingly unreliable. His drinking habit had become worse and as a result I would often step in to cook his breakfast shift - something he would pay me 10 euros to do. I had always enjoyed cooking and as I certainly didn't fancy being a cleaner again (I lacked the self-regulation required for crinkle-free bed sheets) and instead, envisioned myself buttoned up to the chin in a starched, white chefs outfit; metal knife poised above the crispy body of a suckling pig surrounded by clapping guests. With nothing to lose I thus set about fudging my rather empty-looking CV and a short while later sent in a flurry of applications for a variety of summer cooking positions. After a few interviews I met up with an English couple who ran a summer mountain biking chalet and after taking them a sample of chocolate cake (that I had paid the chef 5 euros to make) I received a confirmation phone call later that week and with multiple whoops of joy immediately set about burning my feather duster.

That summer season came about fast and the meadows were soon a riot of thick green grass and spring flowers. I spent the majority of my winter's savings on a shiny new bicycle and embarked on a career as a cook, a chef and a feeder. Waking up to the smell of freshly baked bread that emanated from the corner *boulangerie* I would hurl myself out of bed and immediately set about filling jugs with an assortment of jus d' fruits, toasting croissants and switching on the ancient coffee machine. After arranging thick chunks of ham and Emmental on a plate I would then set the bacon onto crackle mode before getting as much dinner prep out the way as possible; slicing vegetables, marinating meat and invariably boiling vast pots of potatoes for a *tartiflette* before the first bleary eyes wandered down the stairs mumbling something about Red Bull and protein bars. Once breakfast was out the way the guests would then zoom off on expensive bicycles whilst I hurtled around the kitchen finishing off a few last minute details before slipping into my own pair of padded shorts and disappearing off into the kind of vistas most commonly found on Swiss chocolate wrappers.

The beauty of that landscape is impossible to describe, but imagine if you will the iridescent blue of glacial lakes against never-ending fields of copper green meadows and a scented myriad of summertime vegetation offering a delightfully natural comparison to the perfume shops you find in most airports. In the distance the alpine views stretched out for miles into Switzerland and Austria and the sound of clanging cowbells was ever present.

A sense of reckless abandon filled my insides as together with friends we hurtled past locals gathering seasonal fruit; *myrtilles*, multicolored plums, tiny sweet pears, cherries, apricots, apples and fistfuls of wild *menthe*. The abundance of this land, its clear mountain streams and warm sunshine made for some glorious memories and some truly epic fruit crumbles and given that most bike riders are of the XY chromosome variety, there was nothing, absolutely nothing I could have wanted more from life.

The following year saw me earn a few more stripes in the kitchen as a second-in-command to a talented young English chef called Matt and after a much more focused series of applications I managed to line up my third winter season as a private chef in the swanky resort town of Verbier.

Just before that season kicked off I moved into a small flat right off Place centrale with a girl called Katie, one of the most outrageous people I have had the fortune to meet in my short life. Katie was flamboyant; she had trained at a circus, she loved boys, booze and parties and her wardrobe was full of clothes straight from the London theatre school. On several occasions I would bump into her at our flat door with a boy on her arm after an all-nighter at the local pub whilst I was switching on my head torch intent on getting to the top of the nearest peak for sunrise. We would laugh and high five one another in the doorway and the poor chap, suddenly looking nervous, would be forced over the threshold and told to get into the bedroom. I would then have a last minute 'sandwich' check and shout a loud goodbye before clumping out the building straight up the main street past groups of people still lurching around outside the nightclubs. Up, up, up and away I would flee, into the pine forest for a moonlit ascent of the mountain before breakfast duty commenced.

Freedom.

But this particular season differed from the previous years and the pressure that comes from wealthier clientele was evident in the way one was expected to dress, behave and heed every call. I was fully accepting of my lowly status but as the season stretched on I became more sensitive to the underlying current of frustration and boredom in the lives of those that I scrambled eggs for every morning. Most of these guests arrived in the resort on their a once-a-year family holiday but instead of laughter and fond banter there was agitation and a lot of swearing. More than ever before I received complaints and criticism - not just about the food, but about everything

- there isn't enough fresh snow

or it's soooo cold,

there are too many people on the slopes and the lifts are slow,

where is the 'traditional' food?

(Cheese and potatoes) but on no account was I to serve anything too heavy (cheese and potatoes),

what about foie gras? (I refused to buy it)?

Sushi?

More chili, too much chili.

It was endless.

The children were perhaps the worst; spoilt and loathsome, decked out in Versace ski outfits they whined, fought and sulked constantly. Several pairs of skis had inlaid diamonds – fake or real I couldn't tell and I became a great fan of peering through the chalet dustbins to find the price tags of some newly purchased garment that would inevitably end up in the same place by the end of the week.

Every Saturday there would be a fresh influx of guests as the old ones left, more often than not looking at you in the eye for the first time and exclaiming 'How fast the week has gone, gosh, golly!' etc., as their ugly new Range Rover belched fumes all over the doorstep. If you were lucky, they might also slip you a tip but more often than not the real prize lay in what they had discarded, including a one-off jar of caviar complete with a $730 price tag stuck to the lid - I could sell that on eBay! Unfortunately, I never got a chance to realize the outcome of that particular transaction because the very next morning I found the exquisite casing (mother of pearl) on its side and a half eaten baguette wedged behind the sofa after one of Katie's late night mop-ups.

Thanks, mate.

One Saturday in the third week of the season, still getting used to the idea that people will pay hefty sums of money for a stint in a wooden house, I heard the doorbell of the chalet *trrring*. Fixing an endearing smile on my face, I quickly shoved a piece of smoked salmon into my mouth and choked. The salmon dropped into one of my lungs, forcing me to spasm for several minutes as I attempted to Heimlich maneuver myself against the fridge.

The bell rang again, and again. After regaining some composure I headed downstairs and swung open the massive oak door with a humble, welcoming smile pasted on my blotchy face. There in front of me stood the ugliest couple the world has yet put together – the sight alone forced me to reel back in horror. My eyes were spared further assault as a large fur coat was dumped on my face as the offenders simply bombarded my eardrums with the most insulting Russian accent ever uttered by humankind. They were angry and tired and I was promptly shoved aside and given instructions to "BRING BAG."

On removing the coat from my face and dragging their golden suitcases upstairs I escorted the couple into the living room area and graciously offered them a slice of perfectly moistened lemon cake. This was greeted with wide-eyed indignation and a brief moment of silence before the words "VERE IS ROOM FOR DAAG" were shouted with enough force to send me stumbling backwards into the dustbin. My terror was further exacerbated by the confusing reality that no matter how wildly I looked around the room I was unable to locate this mystery *daaag* anywhere and given that all twenty-three en-suite bedrooms where at their disposal I had little choice but to stand there in stunned confusion as my ears dripped blood all over the mohair carpet. Ten minutes later, having secured assistance in the form of an emergency visit from my boss I was fully occupied in attempting to calm the still bellowing couple down and in the process had located the aforementioned *daag* lying silent and still inside the woman's handbag. The *daag* was one of those small rat-like creatures around whose neck was a collar studded with bright red jewels. I wanted to squash it with my foot and put it out of its misery but instead found myself patting its fur-less forehead with my pinky before getting yelled at once again.

Natasha, my boss, swiftly arrived on the scene, looking somewhat agitated as she too suffered verbal abuse accompanied by much gesticulating at the creature in question who continued to lie captive and sad in its Gucci tomb. Leaving her to handle the problem I returned to the stove and set about trying to impress anyone who cared to watch (no one did) with a vigorous whisking technique I had perfected over the past few months.

Forty-five minutes later I was summoned to the dining room for a short debrief of the situation at hand while the Russians glowered at me from the corner. Still trembling a little I kept my gaze at half mast and nodded rapidly whilst I received my instructions to remain in the chalet for the rest of the day and coordinate the arrival of a team of people who would be delivering the required accoutrements for 'room for *daag*', which, when they did finally appear just as the sun dropped behind Mont Blanc included a roll of fresh grass, a small water fountain and three potted ferns.

*No one ate my cake.*

Perhaps the most outrageous event by far, occurred a few weeks after the shouting Russians took leave - just as the snow had begun to acquire the faintest streaks of brown that marked the end of the coldest months. After four months of kissing ass and uneaten chocolate fondants I was thoroughly fed up of dealing with the stream of ungracious nob heads that sulked into my life and began ticking off the days until the summer months and the mountain bikers returned. You can imagine my horror then, when one deliciously crisp morning just as I was wiping the last blobs of hollandaise from the counter I was summoned to the breakfast table by yet another monstrosity and given the instructions to prepare a selection of canapés for 100 people that evening. "IT IS BIRRRTHDAAAY. WANT TO MAKE PARTY."

*Tonight?* Shut up!

Plans of a day careering down the ski slopes with a handsome Scotsman and a whole tub of leftover pork chops dissolved instantly as I looked the beast in the eye and told him it would be my pleasure. Several hours later, with salmon blinis and Parmesan soufflés littering every surface I lay gasping and exhausted in the corner of my kitchen; I had achieved the impossible - the bastards had better appreciate this.

At 8pm the guests arrived; they had been flown in on Mr. Big's private Lear jet and the chalet instantly transformed into a multi-billion dollar fashion display. Diamonds, emeralds, gold teeth and fur coats - I swore one was made of polar bear - and that dreadful accent instructing me to bring more, MORE, NOW! Dozens of flabby white jowls worked their way around my red pepper soup shots and prawn toasties until at long last, drenched in sweat, I sent out the final tray of oysters covered in yoghurt.

I used yoghurt because at 4am I was close to breaking point and had tripped over so many empty magnums of champagne that I didn't think anyone would notice.

Half an hour later after clearing up as much of the kitchen as possible I cautiously made my way across the hallway that at its furthest point put me in full view of the shrieking revelers. Gathering momentum I fixed my eye on the heavy oak door beyond which lay the path to pre-dawn freedom but, turning my head at the very last instant found myself shuddering to a halt as if I had just walked into an invisible wall.

The spectacle that met my eyes was one I shall remember forever for the room I had scrubbed clean that morning had been transformed into a scene of desolation that ISIS would have been proud of. All of the food I had spent the last nineteen hours carefully preparing was now smeared upon every visible surface; cigar stubs floated in murky pools of liquid, antique carpets were piled up together revealing badly scarred mahogany floorboards whilst several pairs of guests clutching bottles of vodka dry humped one another against vintage Buddha statues. I stood there aghast, mouth opening and closing like a chameleon that had just swallowed a particularly large moth and nervously spasmed for several seconds unsure of what to do next.

As thoughts of all the hungry people who could have survived off this fare for several months flew through my brain, my eyes settled upon something else that further increased my incredulity. There in front of me, weaving amongst the thirty-odd guests who had managed to remain upright were a substantial number of dwarfs – perhaps fifteen in total, dressed head-to-toe in black suits with what looked like graduation caps atop their heads. They didn't seem to be engaging in conversation and instead I watched as they slowly circled the room, eyes lowered to navigate the obstacles, hands clasped firmly behind their backs.

Swallowing hard I blinked several times before deciding that either my drink had been spiked or that I was simply too exhausted to see straight so I returned to the kitchen and took a deep drink of water from the tap and put my head inside the freezer for several seconds. I then tiptoed back to the doorway to confirm one or both of the above but to my horror there they were again.

I happened to catch one of the waitress's eyes at that point and summoning her over with a less than discreet two-handed wave I begged her to fill me in. At that point I noticed a few loose tears rolling down her cheeks and thinking she hadn't understood my *whatthehellaretheyherefor* question I helpfully jabbed my finger at the closest one and continued opening and closing my mouth. Looking around somewhat fearfully the exhausted girl then leant in close and whispered in my ear the words, "walking cocaine tables". WALKING COCAINE TABLES!

My eyebrows almost hit the roof, but before we could discuss any further details a large square man resembling a gigantic vole stumbled past and groped my mate from behind ordering her to bring "VODKA, MORE" and so we left it there.

*  *  *

Money then has the ability to make impossible things happen. Money lubricates certain bureaucratic processes like getting a driving license in Kenya and in the very same continent it appears that enough of it can also secure you multiple terms as president. Money also allows you to have a planeload of dwarfs delivered to your front door anywhere in the world and at a more down to earth level it provides one the freedom to eat, sleep, travel and enjoy the smaller things in life.

This realization led to the acknowledgment that although the mountains were a fine place to be - spectacular even, I was no longer satisfied with my current existence and that the time had finally come to push on and find something else.

# Chapter Four

It was Paulo Coelho who wrote the words, "And, when you want something, all the universe conspires in helping you to achieve it." What occurred next proved him to be spot on.

Around the same time as my moral compass was pointing me away from the chalet kitchen I met a man called Dave.

We ended up sitting opposite one another at a friends dinner party one evening just two weeks before the season closed. In between prongfuls of fondue I disclosed to Dave that I had not reapplied for work in the same mountain bubble as I was determined to do something different - something meaningful with my life. Although I didn't know exactly what that was - or with whom it would happen nor where this gloried transformation would take place *I just gotta get out man, before this goddam beautiful place swallows me whole.*

And just like that, a door swung open and Dave - cautiously at first - mentioned an idea that sounded like it could help both of us out in the short term; me with a one way ticket out of the resort and he with a vacancy he needed filling ASAP. Dave ran a travel company that specialized in exchange programs for high school kids and an upcoming trip to Vietnam was lacking an assistant … *Interested?*

And so it was that a few weeks later I found myself standing in the arrival section of Bangkok's international airport surrounded by an estimated *eight billion people.* Fortunately I was able to locate Dave at the pre-identified spot and as we clipped ourselves into our connecting flight to Hanoi, I took note of the emergency exits, breathed in deeply and sent out a mental fist pump to the world for getting me to this point alive.

Arriving at our destination later that afternoon Dave and I made our way along a vast, spaghetti-like collection of roads in a spluttering taxi driven by a man just over four foot tall who garbled animated Vietnamese the entire length of our journey. Every now and again he would twist around in his seat and beam widely, displaying all two of his remaining teeth. The taxi doors were tied together with multiple pieces of string to stop them flying open and as I sat balanced on the edge of the seat, I couldn't help but beam right back.

There were so many people! Like giant columns of Siafu ants moving together along the edge of the road, the majority of them balancing long wooden poles across their shoulders from which hung circular trays filled with plastic toys, scrubbing brushes and branches of hairy red *rambutans*. As we closed in on the city limits, scooters zipped about us like little multi-colored insects until their number reached that of plague proportions; each vessel carrying at least six generations of family members along with the occasional chicken tied to the mud flap.

The streets narrowed further as we crawled into the CBD and the volume of people increased, crushed together by hundreds of tiny shops that spilled their wares into the road. Kids selling peanuts and boiled octopus, pressed their faces to our window and we passed miniature mountains of green coconuts with clusters of old women bent over pots of steaming liquid. Horns beeped incessantly, dust settled and then rose in a never-ending dance, people, fumes, color and chaos - my mountain lungs gasped and shriveled but *phwoaarrr* did it feel good to be back in the real world.

Eventually we pulled up outside our hotel in downtown Hanoi and whilst I untied the door string with sweating fingers a number of beautiful people dressed in silk floated down from the reception area to gather our bags.

Ah ha! geishas! I thought to myself knocking back a welcome thimble of jasmine tea.

Later that day, as the first calming fingers of cool stroked the city streets Dave and I found ourselves back amongst the masses as we fought our way to one particular spot Dave described fondly as his 'local.' After stopping several delighted but hapless passersby for directions we eventually found ourselves wandering down one of the quieter side streets until Dave muttered a triumphant, "Found it!" as we zoned in on a short section of tarpaulin propped up by road signs. A large, bustling lady standing guard at the entrance must have recognized my chaperone for her crinkled face immediately broke out into a wide smile and we were immediately handed two purple cleaning buckets as seats and told to make ourselves comfortable around an upturned cola crate that served as a table. Several seconds later we were both swigging deeply from glasses filled to the brim with something golden and frothy and as a long train of orange-robed monks walked past in silence I fought to contain a large bubble of excitement that rose up from somewhere deep within.

At 5am the next morning the scooter stampede woke me from my twitching slumber and as I drifted in and out of consciousness I couldn't quite believe that at no point in the next ten days would the words "Eggs Benedict" be uttered in my presence.

Even at that time the air was sticky and humid but the sky was lightening quickly and soon enough it was time to join Dave for the much anticipated breakfast rendezvous; a disappointing arrangement of rip-off cornflakes and fluorescent jams that sulked up from the buffet table

Where were my lotus flowers?

The rest of the day passed more successfully however, with multiple meetings, introductions and telephone conversations confirming that in less than twenty-four hours we would be taking sole responsibility for a group of fifty High School teenagers aged between fifteen and seventeen arriving from Hong Kong. Fortunately, any doubts I had over my own abilities were put to rest as Dave introduced me to our two guides: Duc and Quy who would be accompanying us for the duration of the trip. Both were young, energetic, spoke fluent English and were genuinely excited about the kids' arrival and after an enormous bowl of noodles and floating eggs we retired early.

The following afternoon our small team, dressed in matching blue uniforms made our way back to the airport to meet and greet our group as they arrived.

I am often fascinated by the human being and find that airports provide one of the most captivating places to study our species up close. As we had arrived well before the expected arrival time I settled myself close to an air conditioning unit and spent several minutes scanning the crowd for a subject worthy of my attention and before long zoned in on one particular man aged sixty or so whose translucent skin was covered in tattoos of flying dragons and swooping eagles. Despite the humidity this particular gentleman wore a pair of thick army issue trousers matched with a sweat stained vest that covered something lumpy just below his waist that jiggled at the slightest of movements. Could it be? My eyes widened and then creased in delight. Yes! A fanny pack ! God Bless America!

On his arm rested a stunning Vietnamese woman whom I estimated to be about twenty-five, dressed in a flowing red dress and electric blue heels. With her arms clasped around an abnormally large bicep I couldn't help but stroke my chin and ponder the bizarre intricacies of human attraction. A shout from Dave bought me back to real time and glancing up at the arrivals gate I took in the fearful sight of fifty teenagers spewing forth towards baggage claim. Drawing in a deep yogic breath I located some primal sheep dog instincts and swiftly moved in.

Two exhausting hours later, fifty kids and four teachers were finally head counted onto the waiting buses and shortly thereafter we arrived back at lodgings. Once everyone had settled into their rooms I found myself shouting at a giggling, whispering crowd gathered in front of me about the perils of leaving the hotel unaccompanied and the strict no-alcohol policy. Following on from this was a monumental effort to usher small groups of five out of the hotel door where a short walk would take us towards our selected dinner venue for the evening. The fact that anyone made it there alive was a true testament to Duc who took complete control of the situation and led us through a dazzling maze of narrow alleyways and higgledy-piggledy streets with a laid back style that made him an instant hit with the kids.

The restaurant that evening went by the name of KOTO - a not-for-profit establishment whose staff consisted primarily of former street children who received training in catering, hospitality and English. The walls of the one hundred and twenty-seat eatery were covered in photos of smiling youngsters dressed in professional chef whites and as we settled ourselves onto brightly coloured cushions and expansive beanbags I marveled at the legacy of its founder; a Mr. Jimmy Phan whose wide, gracious smile beamed down from several of the frames above us. The food that evening bordered on gastro-magnificence; mountainous bowls of fluffy fried rice, glistening bok-choy covered in soy sauce and chunks of slow roasted garlic, satay chicken pieces and sticky barbecued ribs, hot cashew nuts that exploded in your mouth, succulent ginger prawns, crunchy beef salads and honey glazed duck. Leaning back against the wall on a cloud of sesame oil-induced bliss, I let my mind drift back to Switzerland and reminisced about all the *Raclette* and *tartiflette* dishes that had passed through my hands over the years. I then ploughed on through a bowl of crispy noodles and attempted to bond with my neighbors who were hotly debating the fine details of Justin Bieber's upcoming world tour.

The next few days were frenzied; the kids and their four accompanying teachers were split into two groups along with Duc and Quy as the resident GPS units and the hours flew past quickly as we toured various shops, marketplaces, museums, palaces and the central dam that takes pride of place in the city center. Then there was a slightly bizarre water puppet show and a largely unsuccessful bike ride that no one seemed to enjoy as the two kilometer circuit around an area of flat wetland was deemed quite tiring.

On the third night we all amassed for another evening of aromatic noodles at a small diner opposite the central train station. There was a buzz of excitement in the air and even my city-frazzled nerves registered anticipation in the days that lay ahead, for on close inspection of the neatly printed schedule it appeared that we would shortly be heading North into the mountains where the main 'outreach' part of the trip would take place.

As Dave picked up the pre-ordered tickets from a puckered old man at the ticket counter, Duc, Quy and I did our best to control the kids who were dashing around the small waiting room in pre-overnight train excitement. The majority of them had zoned in on a young teenager selling Pringles, fizzy drinks and chocolate and by the size of his smile it was obvious that the lucky chap would be heading home early that night. What is it with kids these days and junk food? We've just had dinner for Pete's sake ... Concluding that I was turning into a grumpy old bag I did another quick prowl of the room and only raised my voice when enough sticky wrappers had attached themselves to the bottom of my Birkenstocks that it became unsafe to move. "PUT YOUR SHIT IN THE BIN'" I cried - a request that fell on a sea of blank faces finally settled into the waiting seats that were now bobbing along to the invisible beat of a thousand identical iPod's.

As our departure time neared we began the slow hustle of bodies towards their respective 'boys' and 'girls' carriages before locating our own 4-bed compartment mid length down the ancient, wheezing train. Swinging open the tiny door Dave and I entered only to be confronted by two large, middle-aged German women who were busy wiping their underarms with fistfuls of baby wipes. Settling their sights on Dave they smacked their lips with gusto and repeated the phase "Jaaa, gut" several times. Unfortunately for them however, the Euro-orgy would have to wait for both of us were exhausted and as we settled onto our bunks the train uttered a determined *Pweeeeeeee* and we began to creak forwards.

Halfway through a bizarre dream in which I found myself crossing an ocean on a piece of toast I woke up and for an instant couldn't figure out where I was. But the clickety-clack of the carriages broke through the momentary confusion and turning over on to my stomach I attempted to stare out of the tiny window above my head. It was pitch black outside – not a star in the sky, but before long we passed through an area of habitation and the passing blur of a kerosene lamp shed just enough light into its surroundings for me to appreciate that we were deep in the countryside. Sighing gratefully I drew in a deep breath of hot, stale air and settled back onto my pillow as the mechanical rhythm of the train worked their magic on my eyelids.

At 5am there was a sharp knock on the carriage door and loud shouts echoed up and down the carriages. For a moment I thought we were under attack by highway robbers, as did the two Germans who sat bolt upright, their pendulous breasts swaying vigorously beneath frilly nightgowns. But Dave, rubbing his eyes tiredly reassured us that we were in fact nearing our final destination and after clambering my way over a pair of colossal backpacks I headed out down the wobbly corridor to make sure that the rest of our people were awake.

As our yawning, disheveled group sat down to breakfast at a nearby hotel some time later I excused myself and went for a short walk around the streets of LaoCai. We had arrived just after sunrise and I wanted to confirm that we were truly in the mountains; my heart yearned for cool air, tinkling brooks and green space. Unfortunately, it appeared that the Hanoi scooter brigade had followed us there and within twenty seconds I had narrowly escaped being run over three times. I swiftly realized that the mountain wilderness of my dreams was in reality a never-ending row of scruffy hotels and eateries on either side of a road fronted by a myriad of tiny kiosks that sold everything from cheap plastic buckets to cheap plastic bicycles amidst a circus of noise, movement and chaos.

The Chinese border, Dave had informed me lay just over the river and I thought I spotted a slow moving, yellowy snake of water a couple of hundred meters away, but through the dust and exhaust fumes I couldn't be sure.

Suddenly I felt claustrophobic and for the first time since leaving university I registered a flicker of anxiety at the future of mankind. Everywhere I looked there was rubbish, plastic and people.

What was the future of this continent, and of others?
What did all these people eat?
Where did the rubbish go?
And who in God's name buys all those buckets?

I stopped to buy a slice of magnificent looking pineapple from a short, stocky woman perched on a stool in front of a colorful pile of mixed fruit and vegetables. She duly carved up the flesh into bite-sized pieces and wrapped them all up together in a plastic sheet alongside a plastic fork and knife and handed the whole thing over in yet another plastic bag. Horrified at the amount of waste I had just generated on a whim I tried to intervene: "Err… no plastic" (shake finger and head simultaneously). The woman smiled and dropped a plastic spoon into the bag and looked up expectantly. Bugger. No. More finger waving. The woman looked bemused and grabbed a fistful of plastic bags, one of which dropped loose and got caught up in the whirlwind created by a passing bus. "No, no", I stuttered, grasping the sticky bundle of polythene that she waved impatiently in my face. "I …err…. you shouldn't use so much plastic, you know!" Turning to me once more she smiled questioningly and held up a guava. Sod it. Never mind. Thank you very much. Goodbye.

On my way back to the restaurant I noticed a large crowd standing in a dilapidated car park on the other side of the road and unable to help myself, wondered over. Hopping up onto a concrete bollard I peered over the heads of the assembled group and squinting through the dust identified the outline of a man sitting cross-legged in the dirt. Shifting my position a little I noticed that the man wore nothing but a dirty cloth wrapped around his middle and that he held a short wooden flute to his mouth. Unable to hear a sound through the noise of the others I squinting harder and picked up on the silhouette of a large cobra swaying back and forth before him, its thick body, dull in colour and crisscrossed with scars. I have always been fascinated by snakes; mythical beasts of biblical notoriety but somehow the romance of this scene was missing and stepping back away from the crowd I looked up at the polluted, bird-less sky and registered a feeling of intimate sadness.

I arrived back at the restaurant in time for a quick cup of green tea before our group boarded the buses that would take us further North on a five-hour journey into the much-talked-of mountains. Here we would stay for the next seven days in a rural village where the focal point of the trip would take place. Prior to leaving Hong Kong the kids had raised an impressive sum of money that would contribute towards the extension of the local primary school, but what had got me particularly interested was that the primary building material would be made up of discarded plastic bottles collected by the villagers over the past year as opposed to the standard brick and mortar affair. Settling back into my seat next to the driver I did my best to forget the hopelessness of our modern human existence and at the thought of the skills I would learn on such a project and the doors they might open in the future I felt my spirits lift.

The journey was a long one; the road was flat for a while and every now and again put us in sight of the distant shores of China over the surface of the Sông Hong River.

In the distance I caught sight of a few hazy peaks and as the road eventually began to wind its way up across undulating hillsides I drank in the sights of emerald green rice paddies, grazing water buffalos and shrieking bands of skinny children that waved us past in bewildered delight.

The rest of the people we passed along the roadside were busy working their land; they looked fit and lean, their sinewed bodies moving tirelessly beneath the now-familiar conical hat. As I shifted position to try and unstick my sweating thighs from the plastic seat I couldn't help but feel like an intruder as we whooshed past in a cloud of air-conditioned luxury.

I was grateful for the behavior of the kids on that trip, recalling how painful any car journey or school trip longer than ten minutes was for those in charge, but these youngsters appeared quite content to gaze out the window popping bubble gum and dozing peacefully with the constant rustling of crisp packets about the worst I had to contend with.

At long last we approached our destination and clattering down a bumpy dirt track the bus eventually slowed and with much squealing of hot brakes finally came to a halt, disgorging our stiff-legged squad into a large grassy clearing.

Dave had pre-arranged for a gentle hour's walk to 'our' village, an idea that was met with a unified chorus of moans but once we got going the mood lightened and the kids began to poke about at this strange new world of open skies, tinkling brooks and thick green forest. There were hundreds of tiny blue butterflies fluttering about in the sunshine, the occasional shadow of a large bird of pray and once or twice a rice paddy complete with a mud wallowing water buffalo. This last setting in particular created enough of a stir for technology to dominate once again and this ancient, rural scene was captured an estimated thirty million times.

Around about six pm we entered the rural villagescape that was to become our home for the next few days. Mr. Luat, who ran our guesthouse - a simple, bamboo affair that looked out onto the road we had just walked along, welcomed us. With forty-five minutes to spare before our welcome dinner I then headed out into the village proper and took in the sights that can be summarized as three thin dogs, seven snotty children and a few elderly folk who stopped dead in their tracks, stupefied at the sudden appearance of this yellow haired stranger. Pulling into the only spaza shop in the entire village I did a quick scan of its shelves and noted that the local residents must survive off a diet of margarine, soap, potatoes and miniature pots of neon jelly whose expiry date I was fascinated to discover was 2045.

Back at home base I found Dave, Duc and Quy and several prominent village locals, including the chief of Lapantan Village sat around a table talking business. As the only female in this grown up group I felt like a bit of an intruder, but Dave was adamant I join them and a bamboo stool was duly bought forward along with a most unwelcome shot of *rượu*, a local moonshine of toe-curling ferocity. Squeezing my eyes shut I downed it in one go and then sat there burping fire and swaying slightly for the remainder of the meeting until large plastic plates heaped with pad thai appeared and we adjourned for dinner.

The next morning broke clear to the sound of chickens clucking and water buffalos being hurried along the road. After breakfast and a quick inspection of our new work site Dave and I then split the group in half again and we set about preparing the kids for the days ahead. In one corner of the uneven, stony schoolyard was a vast pile of plastic drinking bottles ten meters high and round - an eyesore if ever there was one. Next to that lay the usual assortment of building materials and after a brief demonstration by a team of local laborers we established smaller working units and got busy stuffing the empty bottles with assorted pieces of clean plastic waste (discarded soap, sweet packets, flimsy plastic bags etc.) that had also been collected in vast quantities over time. This process ensured that the bottles retained their shape and once full, their tops were screwed on tightly and the bottles were tied off in bricks of six and stacked inside a supporting frame of wire mesh.

Later on that week, if we succeeded in keeping to our work plan we would later render the walls in concrete and the local children would have a brand new classroom… and a spotlessly clean village.

- Did you know that more than one million plastic bags are used up every minute?
- Or that plastic constitutes 90% of the crap that floats about our oceans – averaging out at 46,000 pieces of un-degradable junk per square mile?

Suddenly it didn't matter that the past few years of my life had been so devoid of meaning because out here I was making amends in helping extend the life of a tiny percentage of this tragic invention whilst at the same time teaching a small group of privileged kids the basic values of recycling.

*I'm actually doing something good for once as opposed to just thinking about it.*

Filled with newfound energy from that moment on the days passed quickly, beginning with an early morning walk before the masses gathered for their breakfast serving of *phō*. Heading out the guesthouse I would stroll out along one of the many dirt tracks that spread out into the bamboo forest, pondering the universe and its mysterious ways and acknowledging the looming reality that in just a few days I would be on my own again with no set schedule or timetable. Being in Vietnam had given me a massive confidence boost and as I had enough money saved up I figured that a logical next step would be to spend some uncluttered time poking about into the whole 'future' business.

Because I was in Asia (which according to the map of the world stored in my head was located smack bang in the center of the world) I decided on the morning of day six that I would make the far distant isles of New Zealand my next destination. As irrational as that might sound at 6am one morning wondering through a forest, it was the only option that I had stumbled upon that filled me with the fizz of adventure and semi-logical opportunity in equal measures.

Here was my reasoning: I was closer to that tiny wee island than I had ever been before, and possibly ever would be.

They spoke English there, which after three years living amongst the Frenchies would be a welcome treat, there were mountains of renowned beauty all over the country and Mum had once mentioned that we had some distant relatives living in Christchurch who were "lovely people."

Buoyed by the confidence of having an idea as well as being halfway there, I put my plan to Dave that afternoon over a plate of fried rice. He smiled, called me a wondering gypsy and confirmed that it was a great choice of destination. That was the final boost I needed and once back in Hanoi I would book my ticket.

As our last full day at the school came to an end the final slaps of concrete were applied to the walls to a lively chorus of whistles and cheers from everyone present. This was shortly followed by a mighty feast held at the chief's house that involved two perfectly roasted sheep and vast, quivering piles of purple rice served on banana leaves. Stretching out on my bamboo bed later that evening I couldn't help but grin into the blackness and utter a final, triumphant *hooooorah*.

*

Once back in Hanoi, showers were used to full effect and the smell of soap, shampoo and body cream filled the hotel, drowning out the ever-present fumes of passing exhaust for several hours. I felt myself shrinking back into my city shell and with half an hour free before we assembled for the last supper I made my way to a neighboring Internet café to book that ticket, afraid that if I waited any longer I might woos out completely.

Upon entering the small, dimly lit café I observed that each seat was occupied by a number of young people playing video games and thus any doubts I may have had over the erratic decision to fly halfway across the world with no set plan was drowned out by the screams and blasts emanating from sixty flashing screens.

Eager to be as efficient as possible I secured a free computer and navigated the various online procedures required to secure my flight and with a final 'click' confirmed the upcoming chapter of my life on a tiny green island far from home. However, any emotion I should have felt -be it nervousness or excitement- was promptly displaced by a fresh round of bone-chilling screams that echoed out from my neighbors computer.

Pushing my seat back I stood up, tut-tutting loudly in an attempt to convey my disapproval at such a hopeless pastime but in the process I couldn't help but peer down at the screen from which came a maximum decibel RAT-A-TAT-TAT of semi automatic gunfire. The graphics of this particular game were jaw droppingly life like, from the tiny hairs on the back of the hand holding the Uzi to the spray of red mist that erupted from anyone caught in the line of fire. Shocked and horrified I picked up my ticket from the printer and fled out the door sparing one last glance around the room full of my peers; the next generation of our species destined to take over and run the world.

What will they become? What future lies in store for them - for us all?
… and what the hell happened to the Mario brothers?

It was once again time to move on.

# CHAPTER FIVE

New Zealand. A distant country that conjures up images of ferocious rugby players and prehistoric landscapes and bands of hobbits feasting in a sunlit glade.

Looking back on my year spent in the South Island my memory settles on the vast stretches of wilderness, the crystal clear waters of its lakes and rivers and the friendliness of the people I met along the way. After the sensory assault of Vietnam, New Zealand stood out as one of the last remaining places on this earth where you could drink water straight from the lake you've just swum in and it was in Christchurch, my home for a short 12 months, where I witnessed the potent combination of fiery young people mixed with infinite opportunity that provided an exhilarating example of what this Change business might actually look like.

*

Shortly after taking off from Bangkok, grateful for my own company once again I sat staring out the window as we passed over the vast Asian continent down below. A never-ending twinkle of urban life that lit up the land for several hours without pause and because I had failed to purchase an inflight meal with my ticket, I was left to contemplate the future of mankind without the welcome distraction of bright orange cheese.

According to my mate Google, Asia at the time had a population of 4.3 billion people, a figure expected to increase to a paltry 5.2 billion over the next 30 odd years. But the figures for my next destination were a lot more pleasant to comprehend. In 2012 New Zealand had a total population of just under four million people, which didn't sound like an awful lot. Nairobi, Kenya's capital city in comparison has a population of three million people; sixty percent of whom live in an ever-widening band of urban slums and as we continue to hurtle towards that nice round figure of 2050 (when the human population is set to hit the scales at 10 billion)' I couldn't help but feel overwhelmed at the thought of what lay ahead.

After what felt like several months worth of connecting flights via Dubai, Tokyo, Singapore, the Maldives and Jerusalem, our plane eventually touched down in Christchurch and stumbling through the airport terminal at what equated to 3 o'clock in the morning, four days ago, I was stopped at the immigration table and asked to remove my shoes - still covered in splotches of concrete, dirt and what could only have been blobs of toxic jelly. Not wanting to cause any problems I did as was instructed and after a apprehensive five minutes the shoes were returned shining like new... epic first impressions! It was a relief to be able to communicate in English and as a smiling woman behind the counter stamped my one-year working visa, it finally dawned on me that I had arrived at my destination, 8,305 miles from home.

Christchurch is the gateway to the South Island and boasts the second largest population in the country after Auckland. It is much like other big cities in the world and the day I arrived it was covered in dense gray cloud, which always makes for an ominous welcome. But spring was in the air according to the whistling cab driver, who dropped me at the front door of my distant relative's house in an attractive residential area lined with flowerbeds. A beaming woman called Paula and her husband Steve ushered me into their lives with a large pot of tea and a breathtaking display of genuine kindness that I soon learned is a characteristic of most New Zealanders. Paula set me up in their spare room and for the next few days I lay about recovering from a rather heavy dose of jet lag. This mostly took place on a carpet at the far end of my room due to a certain ignorance surrounding the electric blanket Paula had fitted to my bed. This strange piece of equipment was a complete novelty and I would wake up an hour or so after going to sleep bathed in sweat, feeling like a partially cooked Yorkshire pudding.

At the time Christchurch was a broken city, still suffering the after-effects of the devastating 2010 earthquakes that had flattened most of the inner city and surrounding suburbs. The roads were twisted and lumpy, the few buildings that remained upright were skewed and heavily cordoned off and the newspapers spoke of little else but the ongoing, multi-billion dollar recovery effort that was besieged with boring insurance matters, government infighting and thousands of disgruntled residents. Both Steve and Paula were not shy in telling me that their hometown was officially ruined and were in the process of packing up their house to join the mass outflow of migrating residents. But with an as of yet unused degree in Disaster Risk Management and a brief but reassuring scour of the local job pages, it was obvious that I had arrived in a city brimming with opportunity for someone like me; full of energy and with nothing to lose.

One of the first purchases I made to facilitate my new life was a shiny green bicycle that gave me the freedom to explore far and wide across the messy tangle of broken streets. But as time passed and the initial shock of devastation wore off my eyes began to pick up on a smattering of informal clues that suggested the long road to recovery had already begun.

At first I saw these signs as individual examples of human creativity; like the myriad of murals and street fresco's that transformed semi-liquidated houses and lifeless city corners into an interconnected orgy of colour and design. Then there were the sporadic spring gardens that had been installed in the center of abandoned car parks and twisted sidewalks complete with brightly painted benches and banjo-wielding scarecrows that offered passers by a refreshing change of scene from the platoon of swinging demolition cranes. On one dusty corner right by the clapboard house I eventually moved into stood a rusted bar fridge, torn from the guts of a crumpled building and turned into a free standing book exchange; its shelves neatly ordered with books and magazines that anyone could help themselves too. There was a myriad of temporary exhibitions too that popped up overnight (such as 10 brightly painted pianos set atop a cascade of building rubble) and my all time favorite discovery; the 'Dance-O-Mat.' This 2x2 metre rubber pad was located just off the only remaining pedestrian walkway through the old city center and consisted of nothing more than an ancient washing machine, four large speakers and a string of flashing fairy lights. Passing by one evening I watched in delight as three office workers clutching well-earned bottles of XXXX beer tossed aside their shoes, popped a $2 coin into the appropriate slot and for a simple iPod connection. With much whirring and flickering of lights the washing machine jumped to life and as the selected song blared out of the speaker system the giggling dancers preceded to throw some shapes on this bizarre public dance floor to the sounds of laughter, wolf whistles and light hearted banter from the surrounding audience.

Slowly but surely, encouraged by the resourcefulness of these creative displays I began to introduce myself to the various urban regeneration initiatives behind their installation, all of which consisted of diverse groups of young folk who had grown bored of the slow bureaucracy that hindered the healing of their broken streets. Manipulating the power of social media to amass willing teams of volunteers those in charge had seized complete control over the creative aspect of Christchurch's recovery and had set about relentlessly pressuring local government to grant them permission to act. Two years down the line they had achieved more than anyone could have imagined - bringing an electric element of fun, life and color back to the wounded city and in doing so had proven that young people have a hell of a lot more to offer society than just acne and mood swings.

Joining up with the various initiatives was easy too – one simply signed up on online and got stuck in and before long I was a regular behind-the-scenes helper at a dazzling assortment of Pop-Up street concerts, fashion shows, art exhibitions and even a 'one night only' cycle-powered cinema. By the time I was made chief fundraiser at the weekly Roller Derby event (a bizarre sport that involved two all-female teams dressed in miniskirts and elbow pads who zoomed around on roller skates, bashing into one another with gusto) I had been on the ground for several weeks and the anxious capitalist inside had begun to mutter. As each job application (found online or through the local newspaper) came back with a big red F circled on the front page (too young, too inexperienced but mostly too foreign) the fragile wall of self-belief I had built up over the past few years began to show signs of weakness. Thus one morning, after a long and life affirming bike ride in the Port Hills I gritted my teeth and on the way back home stopped in at a small café where I had noticed a *'Chef Wanted'* note fluttering in the window.

Fast-forward five days later and I was back behind the stove, my face sparkling beneath a sheen of grill fat. The Beat Street Café was a somewhat alternative establishment run by a group of dreadlocked women with tattoos, beer bellies and a cheek-reddening vocabulary. The building itself had only just survived the earlier tremors and was held together with suspicious amount of wire, duct tape and cable ties. But the food and general atmosphere of the place was eclectic and it attracted a fascinating bunch of people who lined up, often well out the front door, for the famous 'bro breakfast' and football-sized muffins. Arriving for my first morning on duty, my hair clean and brushed, my voice unhelpfully posh I certainly raised a few pierced eyebrows but after swooping in on the much-hated early morning shift, I was embraced with the first of many herb-scented hugs.

Thus my routine began once again. At 5am I would screech down the hill from my house, to the café peddling like mad in the dark, wary of the city and her shadows at that hour. After wrestling with the padlocks with frozen fingers I would then scuttle about behind the cockroaches, switching on the lights, deodorizing the toilet and helping a hovering man called Glen to his seat. Glen was a crooked old man who lived next door and survived off a daily serving of poached eggs on toast. He was a nice enough chap in a rather awkward early-signs-of-dementia kind of way but he never forgot his *pleases* and *thank yous* and for this I lashed his breakfast with enough hollandaise sauce that by the end of the week his sweater had turned yellow. My afternoon hours were free for me to continue my devout volunteering and I would retire to bed early most evenings exhausted, still smelling of burnt sausage but otherwise content.

But as time continued to drift on I could feel myself growing frustrated by my continued lack of application in the real world; quick to snap at the blundering dopes who worked as waitresses and baristas and would vent away for hours in my airless, greasy corner of the kitchen; wildly mixing up the ingredients for a carrot cake to dispel some pent up energy. But the human brain is a hungry beast and given that I was working in an industry besieged in wastefulness and inefficiency it was possible to distract myself with a number of covert 'Jess missions' that included a six-week effort to separate the café's garbage.

My plan had been to reduce the alarming amount of food waste thrown out at the end of the day (on average twelve large plastic bags that would magically disappear every Monday night). Fearing that all those untapped nutrients were being un-ceremonially dumped atop some reeking landfill, I began to contact local pig farmers determined to absolve myself of a small percentage of guilt. Eventually I zoned in on an earthy sounding 'Mr. Brown' who agreed to swing past three times a week but unfortunately (for I have since discovered that pigs love gingerbread) a series of conversations like this one proved too much for my flappability threshold and several days later I admitted defeat.

Me: WHY is there silver foil in the GREEN bin?
Stoned Sheryl: ahh yeeaah, because the peegs don't like that ey?

Me: No, pigs don't eat silver foil. It would stick in their throats, please put this in the red bin.

Stoned Sheryl: ahh sweet yeeaah, who geets the red bin?

Me: Shut up. The garbage man… and WHO put this sausage roll in the GREEN bin?

Stoned Sheryl: Ah no worries, my daad sez peegs laave sausage roll.

Me: You can't feed pigs, pigs, Sheryl!

Stoned Sheryl: Yeaaahhh but there eesn't a bin for sausage rolls.


But in all of this I learned a lesson in humility that was compounded with every dreadlock I fished out of the stockpot and stale sandwich I rescued from the garbage. Although hardly riveting the job guaranteed me a salary and a necessary structure to my day and often, in the quieter times between the breakfast and lunch rushes, I would grudgingly take a moment to reflect on my life and the fortunes liberally sprinkled within it until it dawned on me– slowly at first- that I had been spoiled. I felt like I was owed a 'good' job, it was my right, my privilege but as the days swished past at increasing velocity and my window for 'other' job applications narrowed further a bittersweet reality hit home *it has to come from me.* Somewhere along the line I realized that if I really intended to do something spectacular with my life I was going to have to put in the legwork myself and stop expecting people to pitch up and do it for me.

With this newfound understanding in place I began to search for the appropriate occasion that would allow me to take on a project of my own – to prove that I was more than just a volunteers version of a Jehovah's witness or a disgruntled cook in the back of crusty café. Sending out my latest wish list to the universe I was once again humbled by her responsiveness in a series of events that further opened my eyes to the wonders that abound in our funny little world, but which, at the same time also taught me to be very, very careful for the things I ask for in the future.

A few days after the pig-bin fiasco I found myself cycling down a lumpy street, my feet still steaming from a particularly busy lunch rush. Glancing up my eyes fell upon a twenty foot stretch of wall covered in ugly black 'tags' that read gangsta messages such as *"Bugger off Bianca"* and *"Shit for brains woz here."* The word *'faggit'* in lime green capitals stretched along the entire length of the wall and the closer I stared at it, the more I sensed a feeling of malice and anger radiating outwards and just like that, with an audible 'poof' a light bulb appeared above my head and I realized I had just stumbled into my next assignment.

*Eureka!*

This was my chance – my platform – a one time opportunity to transform this tiny stretch of human hopelessness into something positive, it was perfect; close to the café and opposite an old people's home (think free knitted sweaters and home made biscuits) and that anticipatory tingle suddenly turned into a flood. Flinging my leg over my faithful steed I immediately careered off to the closest paint shop and breathlessly informed the manager and his nubile young assistant of my mission. Twenty minutes later and weighed down beneath multiple tins of sample paint I returned to the wall and toyed with the idea of going through the lengthy process of requesting permission from the local council.

But impatience is one of my many character flaws and besides '*I was a street artist yo.*' Deciding to throw caution to the wind I then set about splashing away with a brand new brush and a tin of sky blue emulsion. Encouraged by the lack of wailing police sirens I began to envisage the final outcome and decided that a large, gulping chameleon with pointed horns, flashing eyes and some uplifting words borrowed from Dr. Seuss would mark my debut into the world of urban redemption.

Over the next few days I made more friends then ever before, ranging from little old ladies who dropped off crust less sandwiches at lunchtime to random dog walkers and an entire demolition team who were busy pulling down the house next door (men in uniform covered in dirt = good). Fuelled by a single-minded intensity to create good in this miniscule part of the world I rediscovered a sense of flow in my day to day existence that kept me buzzing like a 10 watt bulb deep into the night.

One afternoon, hot and covered in yellow dribbles, my muse was interrupted by the image of a dirty man hovering well into the boundaries that marked my personal space. Turning around I did my best to ignore his presence until the briefest of taps on my shoulder forced me to turn around and stare into the fuzzy irises of a guy whose breath smelt of stale beer and unloved gums. *Oh god – please go away.*

After regaining his balance, the man - I never did discover his name - told me that he loved painting and would like to join in. I raised an eyebrow, but before I could go any further he slurred on, telling me that he was part of a team of individuals who had a weekly session of community service (*even better!*) and that if I would accompany him to the police station, he might be able to get permission from his supervisor to bring a few others along too.

Nodding vigorously - anything to remove myself from the stench of his exhale I tried to fob him off with a sandwich. I actually wanted him to go away but he pitched up the next afternoon, and the next, content to sit and sway in the sunshine and just watch. A week later in desperation, I found myself nervously standing behind the counter of the police station with my drunk attaché in tow talking to a broad shouldered officer whose name tag announced him to the world as Brian. Rather surprisingly (for Kenyan policemen would have immediately asked me for a bribe) Brian appeared interested and after several minutes of discussions I scribbled my phone number onto a notepad and was told to wait for his call.

At around the same time as I was putting the final touches to the chameleon I began to receive the odd phone call from individuals who had wandered past and enquired into the possibility of me continuing my work at another location close to their hearts. Standing tall and attempting to emit an air of artistic aloofness I began negotiations and envisioning the slow rise to Banksy-like fame committed myself to the reinvention of another stretch of grubby wall located alongside a railway track a few streets to the south. So excited was I by this new found vocation that I promptly forgot all about my date with law enforcement and quickly got to work preparing the new fresco. A few days later, as I was hurriedly stuffing stale cookies into my backpack ahead of an envisioned afternoon of painting, my phone once again chirruped to life. On the other end was copper Brian who announced – rather excitedly for a policeman I thought, that he had received the official nod to give this project a go. *What project?* I panicked inwardly… *what the hell have I got myself into now?*

Not wanting to dampen anyone's fire, I agreed to meet Brian at the police station and an hour later found myself sitting in what I was sure was an al-Qaida interrogation room surrounded by three of the most attractive 'law enforcement officers' outside of the porn industry. Stunned into silence and resisting the urge to fiddle with their handcuffs I found myself nodding rapidly as the new wall was incorporated into their community service schedule.

Wide-eyed and now slightly less focused on their stiff, shiny batons I began to stutter an apology that this was perhaps all a dreadful mistake. But with a reaffirming pat on my shoulder Brian took the lead and explained that there was no pressure and that I would be provided with all the support I required and not wanting to look weak in front of so many toned abdominals, I agreed to a trial period that involved one particular teenager on his 7th round of community service. This boy, named Tyson had developed a rather nasty habit of spray-painting his 'tag' all about the city (on the sides of parked buses, cars and public buildings) and who they hoped might benefit from a more wholesome approach to public redecoration.

Feeling slightly less apprehensive I agreed to give it a go and half an hour later found myself sat in the front seat of an undercover police car on the way to the house of my troubled protégé. After hurtling around a maze of downtrodden streets that marked the outer suburbs Brian and I eventually pulled up outside a lop-sided, ramshackle house. Knocking firmly on the front door that opened of its own accord, our ears were met with the sound of babies screaming, a woman shouting in the background and what sounded like a pack of dogs tearing apart a piglet. Inside, my eyes slowly adjusted to the gloom and I picked up a scene not so different from the imaginary landfill of my nightmares; empty cardboard boxes lay scattered in amongst piles of soiled clothes and pages of old newspapers, a dim light bulb illuminated a half eaten loaf of bread balanced on a broken chair and a strong smell of mushrooms permeated the airless space. Brian took the lead and calling out louder we made our way towards the sound of voices that came from a nearby room. Stumbling into what appeared to be the kitchen my gaze was met by five rough looking teenagers who were slouched behind the remains of a fast food dinner. In there the smell of old booze, urine and unwashed bodies overtook that of fungus and I suddenly felt painfully exposed in my sunny yellow t-shirt and green Jesus sandals. Brian, who had obviously gotten to know this particular family rather well over the years began to chat openly with the sullen, staring group and after one particularly awkward conversation regarding the whereabouts of their mum - *"she's probably facking the man next door, mate"* one of the younger girls was sent off to collect the infamous Tyson; her brother and my new paint mate. After several minutes the boy appeared in the doorway and slunk into position between his siblings; overweight and as grubby as the others but with a sharpness in his eyes that confirmed I was now well and truly out of my comfort zone. Brian continued to take the lead however (I had suddenly lost my voice) and after just a few minutes and much smirking Tyson agreed to meet us at my new wall in two days time.

Crawling into bed that night I couldn't help but feel a large weight pressing down on my chest that I eventually traced back to my ego. What if Tyson ruins my wall? What if he steals my paint and cocks it all up? What about me? But these emotions were counter-balanced by the idea of this newfound opportunity to step up as a responsible member of a troubled society and see what happened. Once I ventured down this train of thought I was able to convince my imagination that no real harm could come with having a go and with a last minute visual of the policemen sans protection vests I fell into a restless sleep.

*

And suddenly it was Tuesday morning; the big day had arrived and having arranged a day off from the café I was picked up by Bryan in his wicked cool car. To my dismay Tyson was still alive and met my inappropriate 'yo bro' with a dismissive scowl. Making our way to the site, like some horribly dysfunctional family unit, we got busy unloading the paint tins before Brian buggered off, assuring me he would return in three hours to check on our progress and to sign off Tyson's community service sheet.

*Err... thanks, mate.*

After outlining the silhouette of a tree on one section of the wall I left Tyson with overly complicated instructions on how best to fill it in. I then moved on a few feet to his right and as I messed about with some bright orange flowers a largely one-way conversation ensued. After several painful attempts I was able to extract that Tyson was 17 years old and had arrived in Christchurch with his mum and eight siblings 9 months ago. His dad, a notorious gang member had stayed being in Auckland and he hated his new home. He found Christchurch boring and ugly and he roamed the streets at night because he felt like it. After pressing him further to explain why he liked spray painting so much he muttered something I couldn't quite make out but it included the word *cunt*.

The rest of the afternoon passed in silence.

I struggled enormously over the next few days with that kid. He was bored, angry and determined to make a mess of every paint job I sketched out. My vision of teaching the next Dali crumbled rapidly and I got fed up of having to repaint over his work after each session. I struggled to find my own 'flow' and every two minutes I felt the uncontrollable urge to check on Tyson and critique his shoddy work. This only added to the negative flow of energy surrounding the entire project and although I did my best to remain positive, I knew that in Tyson's eyes I was just another self-proclaimed superior: disapproving and controlling. One afternoon, after an extra long shift at the café I stopped off at the wall to take a few progress photos to send to the property owners and discovered that some ignoramus had tagged his name right over the work we had spent the last few weeks tirelessly sloshing together. There is an unwritten rule in the street art world I had learned - you **don't** meddle with someone else's work and I had a sneaking suspicion that my new mate and his gang of hoodie-wearing shits were behind this act of defiance. A few days later, tired and relieved that we were nearing the end of his designated hours I watched in horror as Tyson dropped a tin of red paint all over the sidewalk. Throwing my hands in the air I told him he was a useless little turd and when Brian swooped past in his secret cop car to pick him up for the last time I couldn't help but breathe a sigh of relief. Praying that I wouldn't see either of them ever again I went for a long bike ride in the hills and decided that I simply didn't have the energy to go through anything like that again.

By that stage I had eight weeks left on my visa and my brain had begun its desperate whirring once again. Over the past few months I had fallen for this tiny island and (most) of its people but the immigration process was a battle I didn't have the finances, energy or required skills for and thus deemed it time to once again pack up my belongings and continue searching elsewhere.

Later on that week I received a call from my brother whose own journey of self-discovery was taking place in Thailand where he was managing a small fish farm in the forested north of Phuket. He was due some time off and as it coincided with my resignation from the café we decided that a short road trip around the South Island would be a magnificent way to celebrate the end of this chapter. Boosted by the idea of being with my number one bezzie mate I pushed out a final batch of gluten free cupcakes and hired a car.

Meeting Al at the airport a few days later- his hair bleached blonde from the tropical sunshine- I couldn't have been happier and set about filling our newly acquired chariot with an assortment of tents, mattresses, cooking pots and wetsuits before blazing off into a bright purple sunset.

That trip turned out to be everything we could have dreamt of; tracts of misty rainforest and valleys filled with glacial ice, wild ocean currents and sparkling rivers brimming with wild trout. We lived almost entirely off the land for the duration of the trip, plucking wild spinach from the roadside passes, armfuls of mint, blackberries and mountain thyme and spent several happy mornings diving for green mussels that grew in abundance off uninhabited shorelines. Stretching out our tired limbs each night on warm sand beside a roaring driftwood fire we talked, we laughed and we dreamed - just as all young people should and in this way all the doubts and disappointments accumulated over the past year crumbled away, replaced instead by a growing nugget of anticipation at what lay ahead.

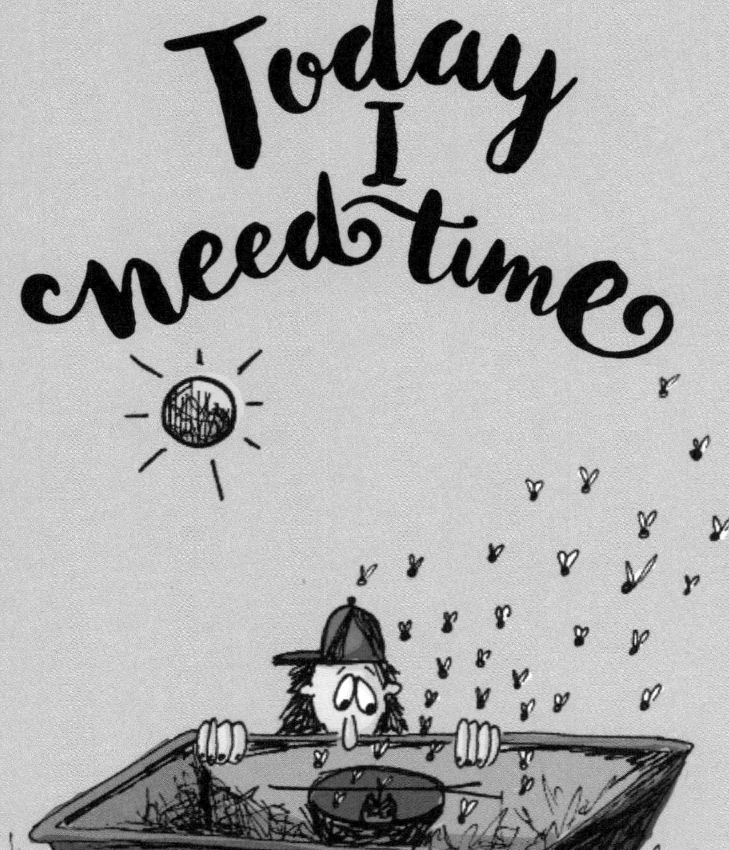
Today I need time

# CHAPTER SIX

Flying due west back over the vast blue Pacific Ocean, I was once again filled with a fluttering of nervous apprehension. But with my brother sat beside me and another year of life experience tucked firmly into my belt I was certain of one thing: no time would be wasted.

Our return ticket routed through Bangkok and thanks to the spirited conversations we had had during our adventures- on the role that young humans could play in making the world a better place, I had decided that a little more time with Al could only be beneficial.

After a brief connecting flight to Phuket, Al's base for the past sixteen months we emerged blinking and stiff into the woolly heat of the tropics. After purchasing a much-needed coconut we then jumped into a backfiring tuk-tuk and spluttered our way to his home; a small wooden cabin located on the northern tip of the island.

My brother had spent the last year and a half managing a small fish farm located a few kilometers over the Sarasin Bridge on the mainland. To get to the site one bumped along a dirt track festooned with puddles and coconut palms before sweeping through a neatly ordered rubber plantation whose trees effortlessly oozed out that curious white sap upon which a multi-billion dollar industry depends. The farm itself consisted of little more than an immense, open-sided warehouse containing multiple concrete tanks dug into the ground and a spaghetti-like assortment of blue piping that burbled with clear salt water sucked from a nearby lagoon, bubbling oxygenators and towering stacks of foul smelling fish food. Together with a two-man team comprising the able-bodied *Lat* and *Hyou,* this hard-working trio had managed to get the place up and running and several of the tanks were filled to capacity with the flashing silhouettes of several thousand Barramundi.

The location of the farm, a pretty grueling job description and a love for the creatures he tended meant that my brother's life was far removed from the bright lights, titty-bars and *doof-doof-doof* clubs that most people associate with Phuket - and Thailand in general. But one late Friday afternoon shortly after we arrived, I persuaded him to take me down to the southern part of the island – to the infamous Kuta Beach, just to see what all the fuss was about.

The whole experience was deeply saddening for what had once been a beautiful stretch of coastline, complete with white sandy beaches, tropical forests and warm ocean currents now heaved with overweight, sunburnt *farang's* on their stag dos. Arriving at our destination we located a seat in the back corner of a seafront noodle stand with a 180 degree view of this lurching tide of humanity and got the equivalent entertainment of front row seats at a Lady Gaga concert. Brutal pop music thumped out from several battered speakers past, which surged thousands of men in white wife-beater vests translucent with sweat, flocks of glittering ladyboys straddling bar stools, touts offered 'special' massage and the elusive call to sexy time and a bustling street market doing a roaring trade in drugs, fake IDs and mango shakes. After an hour or so Al and I had had our fill of humanity and we fled the scene, careful to avoid the droves of UV lit tuk-tuks that swarmed about the road, bringing in the next wave of puking gap year kids.
*How did we make it to the top of the food chain?*

\*\*\*

Al had finished university with a degree in Aquaculture and Agribusiness in 2009 and watching him at work with his team I was filled with a familiar sense of admiration first experienced on the dirty streets of Christchurch.

Accompanying this was a growing appreciation of the role that 'Passion' plays in the lives of those committed to actualizing this business of Change and as I watched Al bending down into his nursery tanks, on the look out for signs of weakness and disease amongst his young fingerlings, I realized that he too was slowly piecing together a role for himself in the future. For in a world where 1 in every 8 people go to bed hungry and 3 billion depend directly on dwindling ocean resources for their livelihoods he had seen a gap – a window of opportunity and a way of combining his love for the ocean and the creatures it contained with an ambitious career choice.

Thanks to Al's patient supervision and a daily exposure to this novel side of the global food production system I felt the studious part of my brain slowly sputter back to life. Memories from my *Food Security* degree began to bubble up to the surface in long forgotten snippets of conversation and hastily crammed revision notes and with guilty admission I realized that after such a long absence from the world of academia, I had turned into a bit of a lemon. Fortunately for me, Al's enthusiasm was infectious and given his frustrations at how far the aquaculture industry had to go in terms of 'sustainability' I felt a tugging urge to slot myself in beside him in an attempt to come up with a solution to mankind's growing demand for protein.

Contrary to what I had imagined, aquaculture or *"the cultivation of aquatic animals and/or aquatic plants for food"* has been around for many hundreds of centuries as demonstrated by early doodles of fish–in-ponds discovered on the walls of several ancient Egyptian tombs. But it is over the last 20 odd years that this particular industry has really flexed its muscles on the global stage and its growth currently outpaces every other form of food production - except that of 'organic'. Nowadays, farmed fish and shellfish make up over 50 percent of the fish sold for human consumption worldwide and that figure can only increase as the availability of wild stocks continues to plummet. Did you know that almost all the marine fish raised in farms are predators such as salmon, sea bass and tuna? These species have the firm, dense flesh preferred by us western folks and as a result, these captive fish are fed smaller wild fish (like anchovies and herring) that together make up the lower section of the oceans food web. It wouldn't be so bad if we ate those little fish directly, but no one I know can pass up a nice fat slab of salmon sashimi and thus the industry does as it's told and continues this frightening game of exploitation until it gets to the stage where it often takes several kilograms of wild fish to produce one kilogram of farmed fish... of which we typically eat only 60%.

It was around this point in my note-taking that I noticed my breathing getting shallower as a pitter-patter of anxiety rippled through my chest like the first rustle of leaves before an afternoon downpour - It had taken me less than 24 hours to realize that this particular industry, just like all the others out there upon which our growling bellies depend had a really, really ugly side.

Added to all of this was the uncomfortable recollection of all those glazed, grilled and battered fish fillets I had served up over the years - a personal insight into how easy it is not to care and as I forced myself to continue probing deeper I couldn't help but sense the bright, tropical sunshine outside darken as if connected to a dimmer switch.

*How can we do this better?*

Given my already suspect background in the food industry and an urge to feel the pulse of purpose once again, I gravitated towards the nutritional side of fish farming - or Aquafeed, as the big brains call it. This was definitely one of the murkier parts of the business; a fact confirmed by Al who had no choice but to spend upwards of seventy percent of his farms annual operating budget importing truck loads of feed from Singapore at an eye-watering US$2.5 a kg. Consisting of small, dried pellets made from the leftovers of fish canning factories and any other creature scraped up off the sea floor it became harder and harder to imagine how the ocean could possibly endure such relentless abuse.

And then, during one morning of melancholic Google-ing I stumbled across a solution; so dazzlingly straightforward in concept and breathtakingly innovative in design that I couldn't help but leap from my seat and engage in an ecstatic sequence of *WOO YEAH'S* around the verandah.

The answer it transpires involves insects.

\* \* \*

In certain pockets of the world the nutritional bonanza stored inside the humble creepy-crawly has long been recognized as part of a well-balanced and seasonal diet. Pliny, the first-century Roman scholar described how Roman aristocrats loved beetle larvae reared on flour and wine whilst Aristotle, the fourth-century Greek philosopher and scientist, wrote of the ideal time to harvest cicadas: "*The larva of the cicada on attaining full size in the ground becomes a nymph; then it tastes best, before the husk is broken. At first the males are better to eat, but after copulation the females, full of white eggs become more so.*" In Ghana, during the spring rains, winged termites are collected and fried, roasted, or made into bread whilst in South Africa the very same insects are ground up and eaten with cornmeal porridge. Continuing this cultural frog-hopping; gourmands in Japan savor aquatic fly larvae sautéed in sugar and soy sauce whilst it is de-winged dragonflies boiled in coconut milk with ginger and garlic that are considered a swoop in Bali.

But despite its long tradition and current favor among at least half of the world's peoples, eating insects is still rare not to mention taboo - especially amongst us coke-swilling, hamburger-munching westerners. Culture plays a vital role in all of this and one suggestion put forward as to why we have acquired such distaste for bugs is that when Europe became agrarian, insects were seen as destroyers of crops rather than a source of food. However most of us will happily knock back a plate of shrimp or lobster (which, like insects, are arthropods) along with pork and oysters, which are food items that other cultures reject as equally dirty and obscene.

Did you know that the average hamburger contains roughly 18 percent protein and 18 percent fat whereas cooked grasshopper contains up to 60 percent protein with just 6 percent fat? Moreover, like fish, insect fatty acids are unsaturated and thus healthier and research has proven time and again that insect farming is far more efficient than cattle production: (45 kilograms of feed produces 4.5 kilograms of beef whilst the same amount of feed yields 20 kilograms of cricket).

The particular organization I had stumbled across - that had resulted in such elation, was one that had capitalized on such impressive facts and figures and which went by the name of AgriProtein. Flicking through their ultra sleek website I discovered that it was run by a small team of South African entrepreneurs (one of whom had authored two books titled: '*The Protein Crunch – Civilisations On The Brink* and *The Story Of The Fly And How It Could Save The World*) and who were currently championing the way forward in an industry called nutrient recycling. According to one short summary on their home page this was described as the process by which fly larvae, fed on abundant waste sources (from abattoirs and organic refuse sites) provided the input to large scale and sustainable natural protein production.

... *Eureka baby*

In order to further explain the intricacies of this miraculous discovery it is important to understand the life cycle of the common house fly, for after several years of extensive research and development it is this much aligned creature that AgriProtein has formed their entire business model upon.

The common house fly *Musca domestica* Linnaeus is one of the oldest and most successful species on earth today. Producing on average one hundred eggs every few days the average life cycle of a fly goes through four identifiable stages (egg- pupae- maggot- fly). Once the mother fly lays her egg (on a nutrient rich nesting site i.e. rotting meat or decomposing organic matter) a translucent, throbbing blob of maggot will emerge from its shell around about day five. Over the next seventy-two hours the maggot will dedicate every waking minute to taking on board sufficient energy to facilitate its final transformation into the buzzing black nuisance we all know and swat and it is also during this time when it is at its most nutritious; containing a mouth-watering array of amino acids, unsaturated lipids, vitamins, minerals and the highly regarded Omega 3 fatty acids.

Using a faultless mélange of nature, science and technology along with several years of trial and error, AgriProtein has come up with a method of harvesting, cleaning and drying the ripened maggots in sufficient quantities to make the procedure commercially viable. Their current capacity of 40 tonnes a day (*a day!*) of maggot-based livestock feed make this one of the most successful examples of Change-Makers-in-action I had ever stumbled across.

At the time it felt as if this was the moment I had been waiting for my whole life: A decisive starting point and a platform from which I could step knowing that never again would I find myself directionless or unfulfilled. From this moment on I whispered into a glass of papaya smoothie *'my future will involve insects.'*

After some further exploration into maggot farming on YouTube and several breathless phone calls to my brother I devised a list of 'maggot breeding' essentials and later that afternoon, in a knee-wobbling display of sibling support I was driven to a large Cash & Carry store to avail myself in the basic materials required for setting up a fly farm. This included: one small glass aquarium, four large blue tubs, four red buckets, string, a fishing net and a selection of bait comprising a chicken carcass, two pork sausages, a fine looking lamb chop and a fish head that Al confirmed once belonged to a young Mackerel.

The following morning I busied myself setting up the goods in Al's expansive garden that stretched a good 50 meters into overgrown scrub that bordered a greasy looking swamp. Outside in the sunshine, beneath the waving coconut fronds and with the occasional sea breeze tickling my cheeks I couldn't help but feel that fly farming was surprisingly good fun and later that evening, once Al had returned from work I took him for a guided tour of my four experimental set-ups. In each of the four blue tubs I had poured enough water to cover the base ensuring that less favorable insects such as ants were unable to access the smaller red bucket placed in the center. Across the mouth of this bucket, I had placed a short stick from which hung the bait – suspended above a shallow sprinkling of bran into which the meat-fattened maggot would drop once they were ready to pupate. Over each of the tubs I had then placed a section of fishing net and a shaded lid - necessary to stop snakes, lizards and any other inquisitive carnivore from gaining access whilst maintaining enough of a gap for the airborne mothers to enter.

The exact sequence of events I was still slightly skeptical of but from my notes I gathered that once the meat had begun to rot (providing perfect conditions for flies to breed) eggs would be laid and thus in theory I would be harvesting my first batch of maggots in as little as 7 days time. After this I was even more uncertain but as I wiped my hands clean of blood and gristle I figured that fate had bought me to this moment for a reason and thus I'd figure it out as I went along.

The next two days passed in painful un-eventfulness for even in the tropical heat the meat took time to acquire that particular odor that translates into fly speak for come hither. Unfortunately in that time period something or someone ran off with both the sausages and pork chop leaving me with just two remaining experiments. Day three got a little more exciting when late one afternoon my brother and I approached the furthest tub, hidden from immediate view by a large flowering bush. As we drew in closer we picked up on the sound of something thrashing about loudly and with the telltale 'thwump' of body-meets-plastic we stopped and gawped at one another as the sound of furious hissing filled the air. Horrified that perhaps we had caught one of the infamous 15-foot pythons, cobras or deadly water snakes that inhabit the area we each picked up a stick and crept forwards cautiously. With loudly beating hearts we rounded the bush and peering out from the side took in a scene of utter desolation: an upturned blue tub, a broken red bucket and a writhing mass of fishing net. Opening and closing our mouths in rapid succession we both uttered an obligatory *WTF* before pulling in a little closer. Spotting a thick, scaled tail and two sturdy legs – each culminating in a large set of glinting claws we identified our captured specimen as one of the largest monitor lizards either of us had ever seen. Although somewhat less frightening than a snake, all 3 meters of the thrashing beast was firmly entangled in the net and given that the Thai's have a unique penchant for lizard stew we figured it too cruel to leave her - him to such a lousy fate. *Shhhushing* stupidly I managed to grab hold of one corner of the net and with much swearing, high pitched yelping and the occasional leap for safety we eventually managed to free the chicken thief from its web. Disappearing without even a backwards glance of thanks our friend vanished towards the swamp and after a brief inspection of the one remaining experiment – the stubbornly fly-free fish head, we retreated back towards the house in dignified silence.

Day 4 and 5 passed in a blurred frenzy of increasingly smelly fish head monitoring – the flies were definitely picking up an interest but with one hand occupied in holding up the lid and the other pinched firmly around my nose I was unable to spot a single maggot – *where were the buggers?*

Day 7 and the entire garden was awash with vast, black clouds of hungry flies – some of which I was certain had come from as far afield as Cambodia. Ushering my exasperated brother out the door as fast as possible I took several large deep breaths and headed out to the tub with a clothes peg in hand and a pair of yellow dishwashing gloves held up above my elbows with rubber bands.

*Today was the day.*

Approaching nervously, already swallowing continuously on account of the smell I positioned the clothes peg on my nose and after one final gulp of fresh-er air groped my way blindly through the dense cloud of insects that surrounded the tub. Waving my hands about like a windmill I managed to clear just enough of a gap to lift up the lid and was instantly enveloped in a warm cloud of reeking ether shortly filled by the mother flies who zoomed about bouncing their dirty little bodies off my eyes, nose and lips…

*Well this sucks.*

Wishing fervently that I hadn't just eaten breakfast I tried to suppress the bile rising from my stomach. The stench was almost unbearable and my hands appeared to have frozen to the lid, unwilling and unable to move closer to what had once been the fish… *but wait… hey look… movement!*

Due to the possibility that you too may have also just consumed breakfast I will keep this next bit short, suffice to say that what little flesh remained of the mackerel now positively heaved with maggots; big ones, small ones, yellow ones and white ones, maggots with eyes, maggots with feet, young ones, old ones, long thin ones and fat, squishy ones –like an insect version of Glastonbury festival just before the headline performance; wriggling, writhing, pulsing and squirming.

And then at that precise instant the smell disappeared, the flies stopped humming and life suddenly acquired a solid sense of meaning. I then experienced one of those rare out-of-body moments where the consciousness takes flight and finds itself floating above the physical being and what I witnessed down below was not a deluded young fool kneeling in front of a bucket, but a visionary. A determined crusader prepared to forgo her ego and the expectations of modern society to break new ground, to try and find a solution to one of the many problems we face in this crazy modern world.

*THIS was it.*

However, shortly thereafter one of the larger, airborne specimens - I believe it was a blue bottle, careered straight into tonsil and with a panicked *UAARGHHHH* I stumbled backwards, retching violently. At that point my consciousness lost its footing on its ethereal balcony and swiftly re-entered its mortal shell with an almost audible *'poof'*. Recovering slightly, my entire body now covered in sweat I gritted my teeth and removing a spoon from my pocket, dove back into the melee and scooped up 10 of the maggots closest to hand before dumping them quickly into an old yoghurt pot, later to be transferred into the glass aquarium.

By the end of the second week Al and I had witnessed the entire life cycle of what we assumed was *Musca domestica* - an interesting series of events indeed, but I will admit that by that stage my enthusiasm had spluttered out almost entirely. Fly farming was hard, repulsive work and in the absence of a microscope and a PhD in animal science, nutrition or chemistry I felt limited by my ability to move on from the tubs in the garden phase. Thus, when Al left for the farm one morning a few days after the final chrysalis had released its airborne charge, I quickly sprayed the whole thing down with doom and packed away the rest of the equipment out of sight beneath the stairs.

At the end of the day, I was still none the wiser on what I wanted to do with my life, but at least the list of 'maybes' had been narrowed down.

# Today I made a mistake

# CHAPTER SEVEN

One day shortly after the Spirulina escapade something miraculous happened that shook me out of the dark, rank place I had settled in and proved once and for all that the universe was watching.

I had gone to bed at my usual seven-thirty desperate to escape the drudgery of daylight hours as fast as possible. By 5am, after a solid nine hours of tossing and turning I was wide awake and fully aware that if I lay still too long, those dark, goading demons would creep up on my subconscious and attack en masse. Yawning loudly I shuffled downstairs, made a cup of tea and opened up my computer. For several whole minutes while my laptop whirred to life and an Internet connection was established I sat staring at the fruit bowl.

My email account always opens first and my eyes drifted unfocused through a number of new messages: Spam, spam, oh looky - another rejection letter. And then, right down at the bottom my eyes zoomed onto an unread mail whose subject line read: Re. Job Enquiry- Successful application– When can you start?

The world stopped; it actually did. After twenty seconds of mute disbelief I pinched myself. Then I made another cup of tea, cut open a pineapple and washed the dishes, paid a visit to the long drop, changed out of my pajamas and walked around the garden, savoring the suspense like a kid on Christmas morning.

Eventually I sat myself down in front of the computer and with much controlled breathing I clicked open the mail. A friend of a friend, who knew someone else in South Africa - in Cape Town, no less - had recently won a big contract for some upcoming event and they needed someone to join their team immediately *blah, blah,* small print, *blah.*

The wave of relief that washed over me was all encompassing. This was it: the break I had longed for, the opportunity to engage with real people in the real world; to prove myself as more than just a drifting, idle youth. God, I'd have a salary, a purpose; a structure to my day.

**I WOULD HAVE A JOB!**

Without any further thought I clicked reply and accepted. No one else was awake yet but I knew mum would lend me the money for the ticket; hell, she'd probably just give it to me in celebration of the fact that I wouldn't be moping around her feet for the rest of eternity. A short while later, once everyone had woken up and drunk their morning tea, I metaphorically exploded; a wave of energy, light and happiness swooped about the house with monsoon-like ferocity and no one could do anything but watch me and giggle in bewildered delight.

My parents, ever the voices of reason, begged me to find out more about this company, the woman in charge, anything, but it took several hours to calm me down first, by which time I had already packed my bags and booked a flight. When finally they did succeed in sitting me down in front of the proffered website, it got stuck loading the first page and only the words 'Luxury' and 'Opulence' were visible in big gold lettering. Not exactly what I wanted, but sod it, really sod it. Saving the world would have to wait - I needed to save myself first.

Two hours later the site eventually opened: lots of partially naked women draped over Lamborghinis, strategically placed diamonds glittering over their naughty bits. Clicking open the attached job description I focused on the very first paragraph:

"An 8-5 office job based in one of the most fabulous cities in the world. We, the company require someone with a keen interest in luxury products and the Art of Opulence and we need YOU!"

The awkward silence from my family at that stage only added to the storm of emotion raging in my brain: anxiety, excitement and trepidation. I'd never had an office job before, I had no smart clothes and I certainly didn't possess the faintest interest in luxury.

Would I be good enough? Has there been a mistake?

I thought Opulence was a bad thing?

At that moment a tiny trickle of guilt dribbled from beneath the blanket of euphoria. Saving the world was never going to be easy - maybe I'd get another break? But then I remembered the empty beehives, the mildew on my spinach and all those blank squares on my calendar and figured the hell was I taking that chance. Besides, I only had another eighty-odd years left on this earth and it was the kids - the NEXT generation who would have to fight for their planet. I tried - I'm over it. Sending a silent apology to David Attenborough I told Mum to boil the kettle for this was the first day of the rest of my life and we had some celebrating to do.

Cape Town

Touch down at Cape Town - *phwoaarrr* do I love that city. Ever since I left university I had wanted to go back. It's something to do with the mountain and sea combination, the surfer boys and the lifestyle culture and the fact that although the roads were smooth and the buildings modern the city somehow manages to retain that comforting tang of Africa; street vendors selling mangos and pears, cigarettes and toffee on every corner, vibrant languages and minibus 'taxis' that stop without warning and disgorge voluptuous mamas in colorful dresses into the middle of the road. In the far distance, vast mountain ranges jiggle in the heat, offering tantalizing views of the uninhabited wilderness, remote hiking trails and opportunities for high altitude skinny-dipping. The memories of friends, adventures and university crushes flooded over me in a giant wave of happy nostalgia.

**I would find happiness here.**

Driving from the airport towards the city bowl one can't help but notice the sprawling mass of shantytown that fans outward from the highway and whose inhabitants no doubt possess less flowery visions of my whimsical Africa. Fortunately, the twelve foot floodlit, stainless steel fence that holds back the writhing masses from the road does a fantastic job of screening off the individual lives behind it and besides, I was far too excited with the prospect of my own life to contemplate dreary things like poverty, rape and starvation.

Due to the urgent start of my shiny new office job I had only one day to sort myself out with accommodation. Trawling the pages of one helpful website I managed to locate a room in a house just twenty minutes walk away from my new office address. This was necessary as having exhausted my meager savings over the past few months, I was in no position to buy a bus pass, let alone a scooter and as luck would have it an old friend offered me the use of her bicycle for the next few weeks.

That very evening I hit the supermarket, which I had established after my spate of globetrotting as one of the finer methods of making oneself feel at home. Purchasing some oranges and apples that would make my room smell like home, I worked on my "hooowzit" and a couple of hours later, as the sun sunk slowly below the ocean I sipped on a chipped mug full of rooibos tea; sleepy but content.

My new housemates were a colorful bunch of individuals. Consisting of a very large man called Dom, aged 65, who slept on the living room couch (as he had done for the past six years), Kobus and Kobus (father and son) who shared a room with their three dogs whose urine one could smell from the corner shop down the road, three super trendy West Africans called Bonjala, Erik and Sammy who worked on the art stalls in Greenmarket square and Helen from Hawaii, the only other female in the house who kindly lent me a spare lock to use on my door at night. Feral cats stalked the barbed wire fence at the back of the compound and I found an old chicken bone under my bed. But with my iPod on full blast and the cracked windows wide open, I was able to give my room a makeover and after hanging a few colorful kangas on the wall, I wriggled into my new bed and lay there twiddling my toes.

*** 

The very first day on the job passed in a haze of introductions, briefings and pre-purchased muffins from Woolworths - reason enough to settle here permanently. I was taken out for a three figure 'welcome' lunch at the V&A waterfront and spent the morning graciously smiling and familiarizing myself with my dashing new colleagues. At 2pm I was plonked in front of a pimping new Apple computer and given a list of basic tasks to do - easy.

On the second, third and fourth day I continued to do exactly that: sit behind a computer and input data, staying late and missing lunch to get it all done, desperately focused on smiling, sitting up straight and not cocking up.

But it was on the evening of the fourth day that I acknowledged the first hint of concern. I had just emerged from the florescent depths of my five-story 'office' mansion in Bantry Bay, one of the most prestigious areas in Cape Town and was traipsing home along the pier. Winter was approaching and the waves were smashing against the sea wall, sending spray high up into the air against a darkening sky. This is normally something I could stand and watch for hours but that evening I just wanted to walk, to move, to stretch my legs after having sat still for the past nine hours. My eyes were sore and scratchy from staring at the screen and my back ached. I still needed to get home and dump my bags, buy dinner and figure out what the hell I would wear tomorrow as I had worn everything in my wardrobe already, most of it twice.

I slept fitfully that night, my head full of the sounds of my friends and family warning me that I was heading towards a slippery slope. When my 5:30 alarm went off I was happy to get out of bed determined to get a dose of fresh air and exercise before another day of inactivity. Blearily pulling on my running shoes and a thick woolen jersey I headed out from my house onto the road that wound its way down towards the icy cold Atlantic.

Hurtling into the surf made for instant brain freeze but as I lurched back out onto the sea, just as the sun cast its first rays above the horizon I was awash with a tingling glow of warmth.

Good Morning, World.

Walking up to the office forty-five minutes later, there was still sand between my toes and my skin stickily tickly from the sea salt. I took a few more deep breaths and embraced the surge of feelings of wellbeing and good fortune that come with participating in the dawning of a new day. Everything about the neighborhood shimmered with extravagance: the trees were shaped to perfection and straightened with posts and the flowerbeds were perfectly symmetrical. Every so often you would get a whiff of chlorine that floated up from the surface of a never-before-swum-in pool and I couldn't help but feel guilty as I scurried past the line of security cameras that followed my movements right to the front door. As I turned into the driveway I caught sight of a small cluster of miniature sunflowers that had poked their heads out between the adjacent walls of a large marble apartment block. They were a wild little group, growing happily in the dappled sunlight of no-man's land in a rebellious tumble of luminous green and yellow. Their presence gave me a further boost and making sure no one was watching, I quickly splashed some water from my bottle over their heads and wished them a lovely day.

Slipping out of my faithful Birkenstocks I shoved my size seven feet into a pair of aesthetically pleasing ballet pumps that gave me sore knees, drew one last deep breath and entered.

Fast-forward forty-five minutes and the last faint wisps of, "Oh, I'm so lucky, life is lovely etc.," had vanished completely for I was staring at a blank Excel spreadsheet once again. My task for the next few days was to fill it in with approximately 9 billion VIP numbers, names, addresses and emails. What?

The same voices that had sabotaged my sleep all night - the ones that had already started to mutter at the closed verandah door that kept the fresh morning sea breeze firmly outside - roared in dismay. In an effort to banish them I stood up and forced a smile.

Would anyone like a cup of tea?

My new colleague, a Polish supermodel-type who didn't smile, ever, nodded and I set off to the kitchen delighted for the opportunity to distance myself from the computer, albeit temporarily.

For the fifth day in a row I gritted my teeth and focused. At ten o'clock, desperate for distraction I nervously crammed my lunch down my throat when no one was watching, spilling a good dollop of mayonnaise into the keyboard as I did so. Every five minutes or so, in a frantic attempt to stay awake I would tear my eyes from the hypnotizing glare of the computer screen and rest them on the great blue yonder that shimmered so far below. Although the sun was hot that day, I shivered from the air conditioning while I tried to figure out how much tea a girl could drink before she collapsed from over-hydration.

And in this way the days continued. The five o'clock alarm was never welcome but without a daily fresh air and exercise fix I knew I wouldn't make it past 9am without wanting to fling open the verandah doors and hurl myself onto the row of glittering Mercedes Benz parked far below. Never before had I been made so aware of the fickle nature of time - Monday to Friday would crawl past as if someone with a shitty sense of humor had pressed the slow mode button on a remote control, while the weekends flashed past as if on fast forward. And oh, the despair of Monday mornings! Just like that first lesson of double maths upon returning to school after the Christmas holidays.

*Work, home, sleep. Repeat.*

I remember the first morning three weeks in when I seriously considered running away and never coming back. The cracks were clearly visible and widening with every email I sent off signed 'Yours Sincerely'. My eyes felt square and my bottom round, and the energy it took each evening to remind myself of the positives of my responsible, grown up lifestyle became harder to find. Yes, I liked having the weekends free; I liked the salary and the flouncy canapés we were served when attending elaborate functions. But most of all I liked 5:30pm, when I eventually left that office and was free to escape into the evening air.

This new sedentary lifestyle had taken its toll on my metabolism and overall wellbeing. My eyes now twitched constantly and my gums were turning yellow. Used to dashing around mountains or at least moving more than the ten steps from the kettle back to the chair and then a further eight steps to the toilet and back ten times a day, I was too stressed to eat and jumped every time the phone rang. My job was tedious, but somehow it worked me up into a lather of sweat; my boss, who I had nicknamed the Kraken, needed my work NOW, as in YESTERDAY, and when I eventually did hand it over, it was wrong or late or both.

But I was determined to hang on a little longer, more afraid of unemployment than the flatness of my current existence.

**Today I'm just going to be average.**

*** 

Until one Tuesday something happened that changed my life irreparably (I've always wanted to say that) and which finally woke me to the creature I had become.

I was now exactly six weeks and two days into the job, and counting down the minutes until I could take my cabbage and couscous salad out on to the balcony for a two-minute lunch break, when the Kraken barked my name.

It transpired that her computer was having trouble connecting to the printer so she forwarded me a document and instructed me to print. I remember being relieved - hoorah, a rare chance to stand up and have a stretch. But when I opened up the folder that all too familiar feeling of doom rose up from within. The document was over six hundred pages long, double-spaced, size 18 font, entitled 'draft copy'. From previous experience, I knew these pages would be glanced over in a matter of seconds and then tossed aside. I had watched the office rats do this several times and had strangled the desire to stand up and preach to them about the perils of deforestation and wasting paper because I hadn't wanted to make a scene. Instead, I consoled myself with a self-pitying 'what difference would I make anyway?'

But this time was different; this time it was me who would be pressing the print button and condemning seven trees from the Amazon to the landfill.

My palms grew sweaty and my mouth grew dry. Flashbacks from Al Gore's *Inconvenient Truth* I had watched last week whizzed through my head like a high-speed advertisement of the future. On the screen the cursor flicked impatiently whilst my brain began to play back my morning walk to the office, reliving the images of the overflowing rubbish bins on every street corner and the hordes of homeless people that buzzed around them like hungry flies. Shift my eyes upwards I stared at the twinkling sea and imagined its sterile, lifeless waters slurping beneath the black hulled tankers and fishing boats, intent on eradicating the last of the fish. My ears acknowledged the air con rumbling mechanically in the background as the vast power stations in the North must have shuddered and trembled with the demand.

Two minutes later my eyes were still clamped shut and no one had uttered a sound. Finally I stood up and taking a deep inward breath I asked whether I would be able to print the document double sided - or at least decrease the font size.

The silence was deafening; in fact it roared.

The Kraken stood up slowly, sighing loudly, and I could see the muscle in her cheek spasm in irritation. Fixing her small kohl-lined eyes somewhere above my head she spat out the following words with such venom that it made the hairs on my neck stand up. "Ach, print it, PRINT IT. Now. In fact, print TWO."

Staring wildly around the office I caught sight of the Polish supermodel watching me with a tight smile on her perfect lips - ah ha, so she did know how! We were now up to fourteen Amazon trees, seven of which were directly linked to me making a scene. Glancing back up and out through the balcony doors I noticed a faint breeze teasing the branches of the tall eucalyptus tree outside and suddenly, experienced a moment of perfect calm.

Leaning forward, I hit the print button.

Twice.

At that same moment something inside me thundered. And then that something curled up and died, just like the paper trees. I was floored by my weakness. Any grounding I had managed to achieve with that morning's sunrise bike ride evaporated leaving a bitterness that spread over my tongue and down my throat like a mistakenly chugged cup of curdled milk.

As I left the office that evening, I walked home a different way just to avoid the sunflowers – for I had failed them too. My brain had gone into some sort of meltdown and as I forced one leg in front of the other, propelling myself away from the office, I couldn't hold the tears inside any more. Above the sound of my ragged breath I swore I could hear a deep hissing whisper of something far more sinister than anything I had experienced before and it was terrifying.

Walking along the pier I saw the surf was angry and frothing, its surface covered in greasy yellow foam that thrashed about the rocks. Inside my head was a cacophony of noise so loud that I was unable to separate thoughts from actions, ideas from words and I was afraid. Out of sheer panic I blundered into the public library desperate for distraction and the calming smell of old books.

Nervously I walked about the aisles, calming my breathing, picking up book after book, magazine after magazine, unable to concentrate on words or even images. After a few minutes I felt calmer and suddenly overcome by a sense of exhaustion I made my way into the kids section and dropped onto a purple beanbag.

An hour later I woke with a jerk. The roaring noise in my ears had disappeared and a few moments later the sight of a small boy wiping his runny nose all over his mum's trousers coaxed a genuine smile to my lips. Like a hermit crab poking his nose out into the air after a particularly rough handling I sat up straighter and breathed.

One, two, three - In.

One, two, three - Out.

During my twitchy sleep no one had bothered me; neither the Greenpeace police, nor the men from the mental institute down the road and the guilt at what I had done in the office had subsided. "Be kind to yourself," my brother would tell me regularly over the phone and I decided to follow his advice. I hadn't eaten my cabbage yet so I decided it was time to go home, have a hot shower and play some yoga music while watching the sun disappear.

Just before exiting the library I found myself drawn to the public notice board and its cacophony of posters and colored bits of paper advertising things like lost cats, piano lessons and astrology courses... small snapshots into the lives of people I would never know. Just as I was about to turn and walk out the door, my eyes settled on a bright yellow page pinned towards the top right hand corner of the board emblazoned with the words: **"Beekeeping Winter School -1st session THIS WEEKEND."**

I smiled again; that was two in a row! Beekeeping... bloody hell, I liked bees! Noticing that the session was indeed scheduled for this upcoming Saturday I pulled out my phone and without any further thought dialed the number provided. I spoke to a man called Robert - the bee teacher - and he sounded nice. In fact, I talked to him for several minutes and before I knew what I was doing, I had booked myself onto the course.

# CHAPTER EIGHT

That Saturday morning I woke at 5am with a lightness in my chest and a long-forgotten tingle in my toes. After a quick bowl of cornflakes I pulled on a big sweater and a head torch and on my bike, set off into the misty darkness towards Cape Town's central train station. Upon arrival I locked my trusty steed to a railing in an empty car park and bought a third class ticket to the distant town of Stellenbosch where the bee course was located.

Unlike the vacant streets I had just zipped along the station was alive with humanity, people on their way to work talking loudly and stocking up for the journey ahead with packets of crisps and chewing gum bought from spirited vendors outcompeting one another for price and attention. Sharp-looking business men bustled past me, along with scruffy daily laborers and the occasional, Tasers-at-the-ready, clusters of policemen. Among those who strode forth with the energy of a new day shuffled others whose slumped shoulders and mute tongues suggested the end of an exhausting night shift. Clinging tightly to my backpack and keeping a wary eye out for pickpockets, I located my platform and squeezed my way into a carriage that was filling up rapidly. By chance I managed to secure half a seat located between the outer thighs of two voluptuous ladies in matching headscarves who eyed my rosy red cheeks and twinkling brown eyes in astonished silence before continuing their conversation in animated Xhosa.

I spent the journey just watching, listening and breathing in the atmosphere of the train and the life contained within its clattering carriage walls. More vendors dashed up and down the central aisle selling bargain items chaotically arranged inside supermarket baskets; individual peppermints and caramel chews, plastic key chains, passport holders and multicoloured pens, each item described in detail at loud volume and accompanied by a ditty "now then, a lolly to make you jolly...?" "A sweetie for your sweetie...?" Suddenly, in through the adjoining doors bounced a man with a small radio on his shoulder and a harmonica, singing his rendition of a *Kwaito* classic complete with a nimble dance move and who, upon completing his performance, continued to skip up and down gathering a few coins from those who felt he had earned it. Following on from the Kwaito guy came another much older man wearing a crumpled gray suit and a loudspeaker unit strapped to his chest. He was doing his best to sell a cardboard box full of gospel DVDs that he boomed would "CHANGE EVERYTHING".

During the time between performances I was able to take in the hundreds of posters that adorned the carriage walls. Barely an inch of paint was visible under the mass of paper that offered:

*"Cheap, easy, quick abortions (R300) Dr. HALMISI – Professional*
*– telephone number"*

*"Dr. Phil doctor - bring back lost lover, penis enlargement, love potions and*
*money troubles"*

*"Weight loss Allert- we removed your fat! Advice and potion (R100) and also*
*specialize in professional abortion (R300)"*

I sat there in spellbound contemplation of this strange new world I had just entered. My solitary existence that had all but consumed me over the past few weeks, the one where only me, my thoughts and I mattered had suddenly burst open, exposing my senses, boundaries and comfort zones to a world that thudded with noise, color and the intricate squashing together of hundreds of lives. I saw myself not as a lost youth, unfulfilled and directionless but as a rich kid wearing comfortable shoes and the only snug jacket in sight able to hold back the sting of icy air that whistled in whenever the carriage doors opened. Watching the lady sat opposite me in a crumpled supermarket uniform, two children wriggling about on her lap fighting off sleep, I couldn't help but feel free.

An hour later the train pulled into Stellenbosch station with a screeching of brakes and along with several hundred others, we spilled out into the frosty morning air. The sun had just risen over the mountains that surrounded the small town like an amphitheater and through the steaming breath of three hundred bodies I could just make out the last tinges of pink sunrise that kissed the highest peaks. Breathless with excitement and cold I allowed myself to be swept forwards by the crowds towards the center of town along a street lined by handsome oak trees. From there I was able to pull out the directions the bee man had sent through earlier that week and fifteen minutes later found myself tripping up the stairs into the community hall where the course would be held.

Robert the bee man and his wife Deborah were there to welcome me along with their six children who were dashing around on the dew covered grass outside the hall with uncontrollable gusto. Helping myself to a fresh cup of Rooibos tea and a plate of Deborah's homemade honey flapjacks we stood around chatting for several minutes as the rest of the group drifted in. Half an hour later I was accompanied by five other students, four of whom were large, middle-aged Afrikaans men with thick necks and khaki shirts and one shy, nervous African guy who later whispered to me over a second helping of flapjacks, that he had been put on the course by his boss.

At 8am, our stomachs sloshing full of nourishment, the six of us were shown into the teaching room where Robert would begin his lesson. We each selected our seats behind small wooden school desks upon which lay a large blue book covered in images of flying bees, a notebook and a pen.

Robert had been a beekeeper for over thirty years and it was evident from the very beginning that he knew his stuff. He was a small, wiry man with a gray beard and bright brown eyes that shone out from behind a pair of old-fashioned spectacles. His clothes were neither shabby nor smart; he was a man of the earth comfortable in his own skin and happy to forgo the pomp and fuss of city fashion. His wife Deborah was petite, dormouse-like even with a thick pair of ancient glasses perched halfway down her nose and waist length gray hair pulled back tightly by a large velvet scrunchie. With six young children requiring constant attention Deborah had mastered the art of multi-tasking: pouring tea, mopping up spilt milk, jiggling the youngest on one hip whilst separating two squabbling boys wrestling over the last piece of shortbread.

Leaving Deborah to reset the tea station and set up for lunch, Robert began with a welcome introduction and a brief overview of the course that comprised a full-day's lesson every Saturday for the next six weeks. During that time we would cover all the theoretical information required to start off as beginner beekeepers as well as gain some hands-on practical experience that would take place in the last two sessions.

Robert then requested that each of us stand up and introduce ourselves to the group, pointing first to the giant of a man who sat the furthest away from me and whose paper name tag read 'Bobby'. It turned out that Bobby had spent his entire life farming ostriches in the Little Karoo along with his wife and three sons. As Bobby was now approaching sixty, he had decided to retire with his wife to a small patch of land they had recently bought outside Stellenbosch and he had signed up for the course with the notion that beekeeping would make an excellent hobby.

Max was the next one along and with great difficulty he extracted his 12-foot legs from beneath the table and stood up, brushing his bald head on the ceiling as he did so. Max became known as the clown of the group, although most of the time he missed his own jokes, blinking his pale blue eyes in confusion whenever he said something that left the rest of us in stitches. Max was a stereotypical Boer farmer and like the other four white guys in the room, English was not his first language. Whenever we adjourned for a tea or lunch break, the four men would inevitably gather together and a loud conversation in rapid fire Afrikaans would begin, with Max and his whirlwind arms creating havoc amidst the piles of delicate china Deborah had so carefully arranged.

During these moments I would seize the opportunity to pick up as many biscuits as my hands could carry and go sit down next to Peter, the silent Xhosa man who told us that he had been born in a small community in the Eastern Cape along with seven brothers and four sisters and who had moved to Cape Town a decade ago. During his introduction Peter stood with his eyes downcast and hands clasped together. In a quiet voice he explained that he currently worked as a groundsman on a local golf course and because his 'bass' wanted to start beekeeping in the surrounding forest he had stepped forward to volunteer his help.

The remaining two men were both middle-aged and one would be forgiven for thinking they were brothers. Hank Van-something was a rectangular shaped individual who wore the same squashed green cap, pale blue jeans and dark navy pullover to every lesson. Hank had just retired from running a small hardware store in a town called Swellendam and had decided to join the course as he wanted to start designing and building his own beehives. His counterpart, Bartje - not related to Hank at all - worked as a technician at a nearby dairy farm. He was also approaching retirement age and having read up on a few articles on the profitable nature of beekeeping, had decided to give it a go.

When it came to my turn I suddenly felt rather shy; I was the only girl in the room, a foreigner and a city dweller to top it all off. Standing in front of these men-of-the-earth types I felt like a bit of a cop-out, but everyone listened respectfully and after I mentioned that my home was in Kenya, the Boer brigade collectively banged their fists on their desks and uttered what I hoped was the word "COOOOL" in Afrikaans.

When Robert asked me why I had chosen to sign up for the course, I fixed my gaze out the open window into the churchyard beyond and told my small audience that I was in the process of trying out many things in life, but ultimately I wanted to work outdoors. Satisfied with my answer and with all the introductions complete Robert then clapped his hand and switched on the projector. The lesson had begun.

*　*　*

Looking back on those first few Saturdays now, I can't help but grin. They involved a lot more sitting still and concentrating, which was something I battled with constantly from the hours of 8am – 5pm Monday to Friday but on the bee course it was different.

In that small dark room located in the bottom part of an enormous continent surrounded by strangers, I couldn't have felt more alive and more anxious to start my journey through the mystical world of the honeybee. The life of these magnificent creatures speak a story of purpose, productivity and collective betterment that left me speechless. Everything about them - both as individuals and the way they worked together as a unified 'whole' filled me with a wonderful lightness in my chest and a newfound appreciation of the fact that it's not only us humans who have spent the last billion years evolving from a blob of jelly.

**Did you know that there exist over 20,000 species of bee in the world including several families that live burrowed in the soil or which build intricate nests made of petals? But my course was not concerned with these varieties so much, for when one is referring to both ancient and modern day 'beekeeping' it is the common honeybee that you are most likely be dealing with: Genus Apis, Species Mellifera.**

I loved the way that sounded as it rolled off my tongue.

According to Robert it is from the understanding of the life history of the honeybee colony that man has been able to domesticate and exploit these creatures to his advantage, a practice that has been in play since the earliest days of modern human existence. Early drawings of bees and honey gatherers adorn the tombs of the Egyptian pharaohs (Cleopatra was famed for bathing her skin in milk and honey) and in cave paintings all over the world one can find pictorial evidence of a wobbly stick man carrying a pot of something on his back surrounded by tiny black specks. Up until the 17<sup>th</sup> century when early Europeans took their livestock - including honeybees to newly discovered countries, the American and Australian continents were entirely without honeybees. However, a few stingless bee varieties did exist and were used by indigenous people like the Maya in Central America and by the Aztec in Mexico to produce small quantities of medicinal honey and which also had a use in warfare as described in one book: "When (the stingless bee) was released from Calabashes, they flew out like smoke and flew in the eyes, noses and mouths of the enemy who threw down their weapons and were then attacked by the men of the town."

**Did you know that bees have their own social health care system? Did you know they have a sense of social responsibility? Did you know that the female worker bee only lives 4-6 weeks because she works herself to death? Did you know that a bee colony works together as a unified 'super organism' with each individual working for the benefit of the whole? Did you know that each colony has its own personality – so just like with humans you can get an angry one, a chilled out one, a lazy one? Did you know that bees dance?**

**Magic.**

My brother introduced me to that word one afternoon after we had gone ocean swimming in Thailand. We had both felt the urge to escape the human world for a moment and carrying nothing but a pair of goggles, we pulled our way out as far as we could to one of the many coral gardens that existed about 2 kilometers offshore. After what felt like ages we finally approached a giant coral head and all thoughts of impending shark attack were forgotten as we slowly circled this silent, waving landscape.

Our breathing slowed down as we floated gently in the current at the mercy of the waves and the life that bubbled beneath them. Keeping one another in sight, we let ourselves drift over this underwater garden, taking in the intricate web of life that played out below. Tiny orange pilot fish, curious at the two strange creatures who had just floated into their world swam right up to our faces before rushing back to their gently swaying sea anemone, small black crabs, their stalk eyes flicking backwards and forwards, claws raised in defiance scuttled about on the sandy bottom, tiny limpets and poker dot starfish clinging to the puckered coral. Parrotfish, lionfish, eels, shells, sand worms and so many other things I had never even imagined existed.

When we finally came back to shore, mouths dry and skin itching from the salt, neither of us said a word, each awed into silent contemplation of that underwater world we had so briefly engaged with. Eventually, some hours later, that silence was broken as we sat side by side staring out at the palm trees beyond Al's verandah that were slowly turning golden in the last trickle of sunlight. Al turned to me with his big blue eyes and wide smile, uttering that one word that I now use only when I experience something so profound, so otherworldly that it defies any efforts of further description.

**Magic.**

***

And so the weeks continued to zip past as they have a habit of doing in life and suddenly I was down to the second last Saturday of the course and the much-anticipated practical session. At 8am on another crisp sunny morning after the usual tray of treats covered in sunbeam stickiness we were each handed a crumpled bee suit and told to get changed. With much faffing, stuck zippers and merriment at the sight of Max, who had somehow managed to put his suit on back-to-front, we were ushered outside and told to gather around the entrance of one of Robert's beehives and observe in silence.

With the smallest puff of smoke from Robert (the smoke acts as a pacifier) the roof of the hive was gently lifted off and placed on the grass near our feet. Six pairs of legs then shuffled forwards and for the very first time we got a glimpse into a living beehive. I was immediately struck by the sound of a million wings beating in unison, the potent waft of honey and wax that floated in through my veil and the sight of a thousand striped creatures scurrying about the comb.

It was another moment that required profound silence and a few seconds to contemplate this spectacular new world that the universe had so obligingly placed in front of me.

Robert had selected this particular beehive from his collection at home as it contained a very docile colony. After several minutes of quiet observation I happened to glance up and noticed that Robert had removed his bulky bee gloves, zipped open his bee veil and was peering over the hive like a proud mother watching her sleeping child. Surprise shortly gave way to envy - it was hard to really see the bees from behind the fuzzy mesh of bee veil and after sidling up to Robert, I asked if perhaps I could do the same. After a moment's hesitation Robert nodded, cautioning me to do so slowly. Forcing a wave of nervousness back down into my stomach I too pulled off my gloves and with a painfully loud *zzzzzip* wriggled my head out of the veil and inched forwards back into the group.

A few minutes later Robert uttered a satisfied "AH HA" and with a deft flick of his hive tool pulled out one of the central frames and pointed to the cluster of bees in the top right corner where seconds before he had spied the Queen marching about on the comb attended by her loyal harem of worker bees. Hunching closer together we leaned in and with a collective held breath tried to decipher the royal empress from the tumbling mass of black and yellow bodies. Spotting her briefly before she disappeared into the darkness I took in her slender abdomen and slightly longer wings and closing my eyes I tried to recall all the information Robert had taught us over the past few weeks...

**There are three castes of bee in a hive: The Queen, The Workers and The Drones.**

**The Queen bee acts as the coordinating heart and soul of the colony and is the only fully developed female in the hive. Her role is that of the Mother. After mating with several drones (male bees), she will remain in the hive for the rest of her life laying many hundreds of eggs (at times up to 3,000 a day). She is also largely responsible for coordinating the activities within the hive through a secretion of special juices - collectively called 'Queen Substance' - that are passed about the hive from one bee to another.**

**The Drones are the male honeybees, large and squar-ish in shape. They make up a very small percentage of the hive's total population and they don't have a sting, pollen collecting apparatus or wax glands. They are incapable of any community work in the hive and their sole function in life is to mate with a virgin queen.**

**Lastly there is the Worker bee, the most commonly occurring member of the colony and the real hero in my eyes. The Worker bee is always female and she is the smallest and most active of the three castes; morphologically equipped to do all the work in the hive.**

**The little worker bee has a long tongue and a honey stomach in which she collects and transports nectar and a hairy body that enables her to pick up pollen, which the bees consume as their protein. She also has a pair of miniature 'pollen packets' located half way down her back legs, which she uses to compact and transport this powdery substance**

during flight. **The worker bee also has a series of wax glands located on her abdomen that secrete wax for comb building and she has a very efficient barbed sting that she will use to defend her family - and in doing so, will give up her life**.

Not bad for a chick, eh?

\*\*\*

After the last of my irritating "oohs" had faded into the wind Robert then replaced the frame back in the hive and pulled out yet another in the hope of spotting a newly born bee chewing its way out of one of the brood chambers. Instead however, what greeted our eyes was a fist-sized ball of bees that hung off one of the central wires that ran through the frame. From the slides we had been shown in the classroom I was able to identify this as 'comb building' – the highly coordinated activity that would eventually result in those perfect hexagonal cups we all know and associate with honeycomb.

Just as Robert was about to replace the frame back into the hive he looked up and caught my eye with a mischievous wink. He then told me to hold out my bare hands and as I obliged he slowly lowered the frame with its swinging bee cluster right into my outstretched palms, urging me to keep calm and still. Stifling the urge to scream I stood there, too nervous to breathe and watched the little creatures detangle themselves from their huddle and scuttle about on my hands, antennae working overtime.

The world stopped again for me right then. The cars in the distance fell silent, the breeze stopped blowing and suddenly we were alone in the universe, just the bees and I.

And then just like that something clicked.

**I was going to be a beekeeper.**

# CHAPTER NINE

That night, once I had pedaled home from the station, I paced around the house like a caged tiger, unable to sit still for even a second. I couldn't possibly become a beekeeper... or could I?

*How does one even start that?*

*How would I earn money?*

*Where would I live?*

Sunday morning was worse, but try as I might, that growing seed of recklessness just wouldn't go away. In fact, it seemed to grow bigger and bigger by the minute.

In a desperate attempt to escape my own head I took a stroll over Table Mountain, giving my mind permission to run free amongst the miracles of nature that flourished high up in the swirling mists. Transfixed and more than a little wired from the lack of sleep I drifted along the various walking tracks occasionally diving off into the scrub to reach a particularly splendid viewpoint, allowing my eyes a moment to rest on the huge, wild continent stretched out so far below.

A short while later, as the afternoon colors began to turn the world golden, I found myself leaning up against a smooth granite boulder perched on the edge of a four hundred foot cliff that swept down through a tumble of prickly, velvet Proteas towards the pristine beaches of Camps Bay. Stretching out my aching feet I leant back, nurturing a sense of deep contentment at the world and my life within it, too tired to do anything more than breathe deeply and wiggle my toes in an attempt to stay awake.

Before long my ears picked up a faint buzzing noise and glancing up to my left – I spotted them. Six or seven of my new found friends buzzing about a cluster of tiny purple flowers that smelt of bubblegum. Crawling closer on hands and knees, I giggled for no other reason than that I was overjoyed at seeing those ladies go about their work in such a magnificent setting.

Drinking in another moment of profound connection, I finished the last of my Mars Bar and told myself that this was the sign I had been waiting for - I will now quit my job to become a beekeeper - before realizing abruptly that it was getting dark and I was about to miss the last cable car down the mountain. Whispering a frantic goodbye to my mentors, I took off at a sprint to join the jostling flocks of overweight Germans in the queue at the lift station.

My descent back home from the cable station via a perilous bike ride down Kloof Nek road was remarkably life affirming and by the time I reached the front door, the sky was lit with a million stars that flickered in and out of focus with each breath I expelled into the crisp night air.

Hanging up my jacket, I took a few seconds to steady my racing heart before, quietly as a mouse, making my way upstairs, praying desperately to avoid any interaction with other humans.

Fortunately I managed to avoid detection, but before I was able to celebrate I stumbled upon a trick card that was lying in wait, eager to pop that delicious bubble of possibility I had nurtured all day for resting on top of the soup-stained counter was the latest Economist magazine, its front cover emblazoned with the words:

'Generation Jobless – the global rise of youth unemployment'

' Poof '


Circling the magazine cautiously I tried to tell myself that I was tired and likely to overreact to its contents but I needed to read it - I had to. After several cups of sweet tea and a warm shower I settled into bed, switched on the bedside light and opened up the article, guaranteeing myself yet another sleepless night.

According to the author of that particular article if I were to pack in my job at the office and start out on this new adventure as an inexperienced, amateur bee enthusiast, I would have knowingly and purposely joined the 26% of South Africa's 'Unemployed Youth.' Blow this out a bit and focus in on global youth unemployment figures and it would appear that I would be willingly joining the ranks, the seething desperate masses known as 'Generation Jobless,' along with 300 million comrades.

…. Almost a third of the world's population aged between 15 and 35. Oh.

*

Unemployment is like a visit to the hospital in that even before one steps through the entrance the senses have begun to anticipate the experience: those endless sterile corridors filled with whispering nurses clutching clipboards, the assault of antiseptic aromas and the hushed undertones of anxious visitors. Your heart quickens, you might experience a shortening of breath and you can't help but give thought to the fragile perfection contained inside each human body.

Looming joblessness is similar in a sense that I have already begun to concede that familiar cocktail of emotions that comes from being *sans job, sans purpose, sans income.* The unease that comes with watching your bank balance steadily dwindle, the awkwardness of your reply to the question, 'So what are you up to at the moment?" and the desperation that quickly builds as your days morph into one long stretch of job applications. Ultimately, just like the final verdict from the doctor, comes that moment of crystal clear realization that actually, one just needs to go out and get a job, **any job.**

Square One, here I come again.

And then it was Monday. A cold drizzle had set in and having eventually drifted to sleep just as the birds were beginning to twitter outside, I was brutally awoken by the wailing of my 5am alarm.

Reaching for the 'Off' button I lay there silently staring into space, my mind a confused tumble of exhausted thoughts and emotions. *Who am I-where am I-what day is it-what am I doing with my life?*

After ten hopeless minutes of tossing and turning I couldn't take it any more and staggering out of bed, I pulled on my swimming costume and made my way out the front door into the icy morning air. Immediately invigorated I shuffled back along the road, forcing air into lungs that burned form the injustice of it all and fifteen minutes later, plunged headfirst into the angry Atlantic surf that was pounding the shoreline of Clifton 3rd beach with giant gray fists.

Several seconds later I exploded back out of the sea, my skin and brain on fire and after a series of desperate star jumps to get my blood pumping again, redressed and scurried back to the house to prepare myself for another week at Mordor.

After a wonderful hot shower that left my frozen toes aching, I sat down in front of a bowl of Weetabix and spent several minutes prodding it with a spoon. Drawing inspiration from the tradition of reading one's destiny in wet tea leaves I focused intensely at the milky mash, searching desperately for an answer, a clue, a sign, *anything.*

Sighing raggedly, I made for the kettle, anxious to avoid the consequences of such thoughts this early on in the week. But this time around I felt a ripple of stubbornness sweep through my chest and standing up, I locked eyes with a mangy cat that was mewling away outside the kitchen window and whispered triumphantly,

*I want to do something great with my life.*

*I want to feel alive.*

*I want to be happy.*

Up in the office I fidgeted madly from the very first minute I sat down, grinning like a lunatic at anyone who bothered to catch my eye. When 12pm finally came round I strolled out onto the balcony and breathed in deep that fine sea breeze. The rainclouds had been shooed off elsewhere by a warm southeasterly wind and the scent of fresh mountain Fynbos danced in the air. Looking around, my gaze fell onto the only sign of greenery up there amongst the lifeless concrete: a crooked little lemon tree that lived its life inside a small plastic pot. Strolling closer to breathe in the aroma, I suddenly saw them again: "Hey, look, cool - bees!" zooming in and out of the heavily scented blossom, their wings an iridescent blur, unwilling to stop for even a second.

Instantly I travelled back to one of the very first Saturday bee lessons where Robert had informed us that the collection of nectar, the basic source of energy for the colony is an instinctive and vital task for the honeybee. When a bee alights on a flower, the proboscis (tongue) is extended and inserted into that part of the flower containing the nectaries. The nectar is then sucked up and passed from the proboscis along the oesophagus into the thin walled crop or honey stomach, which will expand greatly until the bee can take no more and is forced to return to the hive using the sun as her navigation guide. During her flight the nectar sploshes about in her stomach mixing with enzymes that mark the first stage in the transformation of natural sugars into the glorious, livings substance you and I know as honey.

Mother Nature - you rock, man.

Abandoning my lunchbox on the railing I moved in a little closer, my nose now inches away from the electric fuzz of striped bodies and I took the opportunity to philosophize over this timeless scene: a symbolic metaphor of natural symphony perfected over the millennia that encompasses the extraordinary bounty of our world and serves to highlight the fragile interdependency that each living creature has with one another.

*I'm a part of that too I whispered to myself.*

Suddenly one of the bees took off from its petal launch pad, swerving around my head at the last minute before taking a sharp right turn towards a clump of eucalyptus in the distance. Jerking backwards momentarily I rushed over to the balcony trying to keep her in sight but within seconds lost her miniature body against the mottled background of the surrounding neighborhood. Downwind of the lemon tree I caught another whiff of its blossom and returning to its spindly branches for one final sniff I grabbed my untouched lunchbox and sighing happily make my way back into the gloomy cavern of the building to collect a fork.

Upon entering the office I notice immediately that something awful had happened. The usual bustle of people and papers had been replaced by a deathly silence - no one was moving. Feeling like a kid who has just been caught at the booze cabinet I glanced up and around the room and found myself staring into the wide eyes of several colleagues who had just watched my entire 20-minute performance with the lemon tree. Instantly covered in a clammy layer of cold sweat, I gulped nervously and scurried back to my desk.

Several long seconds passed in silence during which time I kept my eyes firmly shut, fervently wishing that I could disappear into my keyboard like an ant. From the distance I picked up the echoing clip-clop of Jimmy Choo's on varnished floor and with a bolt of panic cum nausea realized that the noise was growing steadily closer - like the *swoooosh* of an incoming death train. Forcing my eyes open again I squinted up at the approaching silhouette of the Kraken; my body shifting simultaneously into brace position like a half-trained puppy who hadn't quite made it outside in time for the toilet. As she drew in closer my brain radar-ed over the features on display recognizing that punishment was imminent; her shoulders were squared, her face pinched and her eyes dark and poisonous and fixed on mine with such intensity that I feared bursting into flame. She was dressed in a spectacularly tight black outfit that day and like a panther closing in on its kill I was able to sense the outcome of pain and possibly death long before my ears were able to pick up on the strange hissing noise that emerged from her glittering, pinched lips.

My limbs felt wooden and I now possessed visible circles of sweat beneath my armpits but somehow I managed to turn my swivel chair into a position where I could face her- woman to woman. At 10 centimeters she stopped, steadying herself for the final lunge at my jugular before dropping her head so that it lay parallel with mine, enveloping me in a cloud of noxious vanilla bean perfume in the process. Swallowing what could well have been my final swallow I fixed my sight on the thin arch of her over-plucked eyebrows and forced a small, desperate smile.

"WHAT exactly do yuu think yoor doing?"

… "Err, I was having my lunch outside. In the fresh air."

"What were you doing with the lemon tree?"

…"Watching a bee."

Silence.

…"Did you know that bees dance"?

"Git back to work."

…"I'm being serious, bees dance! Who here knew that?"

"Jussus Jiss, don't waste my time over and above what you do on an hourly basis anywhaay. Git back to work or git out entirely."

…."Sorry, so sorry. Oh man, sorry"

But I was trying so hard not to smile that I thought my face might crack in half. My adrenaline had finally kicked in and I wasn't sure how much longer I would be able to hold myself together. Fortunately at that point she straightened and with a final look of loathing she turned and walked away.

Stunned, I remained in position, unable to believe I had survived. Several seconds later, still overcome with a feeling of magnificent invincibility I pulled myself out of the luxurious leather chair and reached over to switch off the computer.

**I'm going to be a beekeeper.**

By that stage the Kraken was talking to the lady in the next-door cubicle so I announced that I was leaving to her bottom. She turned, staring at me with a look somewhere between astonishment and beetroot-tinted rage and then slowly she raised her arm and pointed towards the door.

# Chapter Ten

I sped away from the office on a road that felt as though it were paved with bubbles; straight past the sunflowers whose bright yellow heads had turned brown and droopy over the last few weeks and along the street festooned with security cameras for the very last time.

Reaching the main road that ran along the Atlantic Seaboard like a pale gray ribbon I stopped off at a local supermarket and celebrated my newfound freedom with a bag of salted cashew nuts and a chocolate ice cream. A light drizzle had come in off the sea but the sun was putting up a fight, reflecting off the puddles on the tarmac and making the whole street sparkle.

I didn't feel like going back to the house and chose to peddle my way over to the library instead where I scoured the ancient bookshelves for titles on beekeeping in Africa. I then carefully placed these inside my bag, removing the crap office shoes I had hastily shoved inside as I tore out the office and discreetly handed them to the security guard who had watched over my bike while I had been inside. He beamed a brilliant white smile, "Oh *eish,* they will fit my daughter perfectly, God Bless You, Madam!" Flooded with good intentions and the promise of a brand new beginning, I set off to the beach for another swim.

That evening once I got home I narrowly avoided being trapped in the kitchen by Kobus senior (who enjoyed cornering anyone he could into the space between the fridge and pantry and shouting at them about his memories of 'The Great War'). Fortunately, I managed to leap over his walking stick at the very last second, buttered toast in hand before sprinting down the corridor and locking myself in my room.

When I was sure I hadn't been followed I collapsed on my bed, praying fervently for a few moments alone with my thoughts before the evening television chorus boomed up from below.

For several seconds I stared at the ceiling and then, feeling the urge to spread the big news phoned Mum and Dad, who received my confession as a 'quitter' with jubilation. Dad laughed hysterically and told me he had won a bet on how long I would last and Mum shrieked with excitement, "Oh, Jess darling, I'm sorry things didn't work out. When are you coming home? I'll make you a carrot cake!"
Smiling fondly at their distant voices I told them that I hadn't even begun to think about the 'what next' part, but would call them as soon as I had come up with a few viable options. Then, suddenly overcome with exhaustion, I bade them both good evening and crawled beneath the duvet.

The next morning at 9 after a deep, dreamless sleep, I woke with a start and immediately panicked. "I'm going to be late!" Just as I had flung open the doors to my lopsided wardrobe, a flood of relief washed over me with the realization that in fact, never again would I have to squeeze my arms into a totally unpractical suit jacket. Mumbling an ecstatic "Hallelujah!" at the large daddy long legs who lived on my windowsill, I settled back into bed with an apple and one of my new library books.

Over the next few days I spent my time wandering far and wide, rediscovering parts of Cape Town I hadn't had a chance to revisit since arriving all those months ago. I spent a glorious morning walking around Kirstenbosch Botanical Gardens, which continued to simmer with floral energy and bird song. Lying on the emerald grass staring up at the multiple waterfalls that floated off the side of Table Mountain I dozed peacefully, dreamt vividly and smiled often.

The following afternoon I hopped on the train and went for a long stroll along Muizenberg beach that curved northwards along the seashore like a giant golden sickle, picking up shells and washed up sea creatures reminiscing about the good old days at university where much laughter was shared with friends way up high on the distant mountain tops whose peaks I could just about make out through the spray of tumbling ocean.

I also discovered several open-air farmers markets in a part of town called Woodstock that had recently gone through a posh facelift. Wading my way through volumes of home-cured biltong, exotic chutneys, vibrant green pestos and strange sticks of fried dough called *Koeksisters* I met a local beekeeper called Fred who kept an assortment of hives down in the Cape Point Nature Reserve and who wished me luck as he wrapped up a small pot of Fynbos honey in a brown paper bag.

I would miss this gorgeous city, its fiery ocean and fine white beaches, the mountains and the seafront promenade with its ice cream stalls and evening joggers, but the reality of my new situation meant I was without an income or a valid reason to extend my work permit in South Africa. The logical next step would therefore be to hand in my notice to Kobus and head back to Kenya to start afresh from the comforts of home.

I half expected to feel some sort of apprehension at this rather sudden change of direction but instead all I felt was an overriding sense of contentment at what I had done and an urgent desire to just get cracking. I had finally found something I could do with my life, something tangible, something that mattered and somehow, even in the evenings exhausted from a day spent outside in the fresh air, I was convinced I'd make it work.

And then before I knew it Saturday morning had arrived and at 6 I found myself staring out of the train window at the blurred fuzz of passing grapevines. Today was the very last bee lesson and although I would be on my own from here onwards, I was looking forward to the liberation that would accompany my official beekeeping certificate.

Arriving at the entrance to our classroom that by now felt more like a home than a whitewashed community hall I embraced my fellow comrades warmly and after a final overindulgence in Deborah's honey scones, we settled down behind our desks for Robert's final lesson. According to my notes, this session would be dedicated to the topic of 'Beekeeping in Africa - the Potential'; concrete proof that the mighty Universe remained one step ahead of the game.

\*\*\*

Over the past few weeks as our collective 'bee brains' had expanded it became more and more obvious that the global honeybee industry was in a bit of state. Opportunity and challenge existed in almost equal quantities and every new article I read or note I scribbled down from the blackboard seemed to confirm one side or the other. This was further reinforced by Robert who began the lesson by informing us that the fairytale world that we had so recently stepped into; the one bubbling with magic and mystery, dancing bees and pollen packets was one equally full of chaos, confusion and the threat of darkening times ahead.

There was something deeply sinister at play in the world of the bee with reports of vanishing populations across the globe and much confusion as to where they were going, why and what the implications would be to life on earth. It was Albert Einstein who once famously stated, "If the bee disappeared off the surface of the globe, man would have only four years left to live," but at the time his warnings were dismissed as exaggerated and apocalyptic.

Since then however, the FAO has released several reports estimating that out of **some 100-crop species that provide 90% of food worldwide, 71 are bee-pollinated. Bees are essential for things like nuts, melons, berries, citrus fruits, apples, onions, broccoli, cabbage, sprouts, courgettes,**

**peppers, aubergines, avocados, cucumbers, coconuts, tomatoes and broad beans, as well as coffee and cocoa.**

Although the above listed items are not considered staples like wheat and corn (which are largely wind pollinated) they are amongst the fastest growing and most valuable parts of the global farm economy.

In summary therefore: **We. Need. Bees.**

A few decades ago in a few places dotted around the States and Europe whispers began to circulate of local bee populations indicating signs of distress. An alarming number of bees weren't surviving so well after the winter months and entire colonies would suddenly 'disappear' without a trace, a phenomenon later given the name Colony Collapse Disorder or CCD. Fast-forward to 2014 and this is now confirmed reality and front-page news as people from all corners of the globe throw up their hands in anguish claiming that their bees have vanished.

The reasons behind this chilling disappearance is a complicated one, often made more so by the efforts of the big boys of the agro-chemical world - the Monsanto's, Bayer's and Syngenta's whose multi-billion dollar companies have the budget for big PR 'it-isn't-us' campaigns. But no matter how loudly or emphatically these corporate giants proclaim their innocence, be cautious, be wary, for there is often wickedness at play.

"Systemic insecticides - such as those used as seed coatings - which migrate from the roots through the entire plant, all the way to the flowers, can cause toxic chronic exposure to non-target pollinators, including bees. Laboratory studies have shown that such chemicals (often lumped together under the name neonicotinoids) can cause losses of sense of direction, impair memory and brain metabolism and cause mortality."

But there are other factors at play too, including both pre-existing and newly developed strains of diseases and the spread of invasive species like the

Varroa destructor mite, plus the usual culprits of habitat loss, reduced foraging cover, air pollution, shifting climate belts and good old human ignorance. Wrap these all up together and suddenly it's not only Houston who has a problem.

Bees make fantastic bio-indicators, their livelihoods much like our own are dependent on healthy, functioning ecosystems where things like plants, soil, sunlight and rainfall all mix together in an intricate cycle perfected by Mother Nature over the millennia. Thus, when bees start dying, disappearing or getting sick - and no one can work out why then it is vital that we change our course accordingly, for should the metaphorical miner turn the corner and find a dead canary on the floor in front of him - it is a sign that his time is also running out.

**The bees are trying to tell us something.**

\* \* \*

As the afternoon sun began to cast its shadows against the back wall of the classroom Robert wrapped up our course with one final message. He told us that being a good, honest beekeeper in these crazy, modern times would not be an easy path to follow.

"The work is long and arduous, the returns are limited if you don't know what you're doing and you will spend hours working well outside the usual 9-5 window in every type of weather. Then there are the problems of the global bee industry and we haven't even touched on the poison contained in the stings that can build up in the body over time ..."

But at that point I switched off and closed up my notebook. Five minutes later we were released from the classroom and I bade a final farewell to my friends before setting off for the station with shoulders firmly squared and my head held high.

For just as I had first discovered at university and since witnessed in different parts of the world, there were always ways to 'do' the future better. Thanks to the bees I now had a focused start point and a solid understanding that nothing in life worth doing will come easy.

**Ready or not, here I come.**

# Chapter Eleven

*How does one actually begin a new life?*

With Mum and Dad as my enthusiastic support team, I jumped wholeheartedly into the world of beginner beekeeping and after a few weeks of phoning relevant people, visiting existing projects and reading every bee related article I could lay my hands on, a small door began to creak open beyond which lay a world shimmering with untapped possibility.

Thus with a little foresight I can now answer that question with ease:

**You simply begin.**

It took time of course, a couple of months at least to find my feet and generate momentum but I quickly discovered that the more energy one puts into something the more one can expect in return.

Suddenly everywhere I turned there were bees or bee related things; on posters, in books, the Internet; courses, newsletters, petitions, forums and workshops. Everyone I spoke to seemed to know of someone or something bee-related and after several weeks the word got around that I was a beekeeper and my phone began to chirrup. There weren't so many bee folk in this part of the world you see and people are quizzical creatures, many of whom required advice on their own hives or other such bee related assistance.

Each morning I would wake up with the birds, have breakfast and then wander down through the wet grass to visit one of my very own beehives that had finally acquired some tenants during the Cape Town chapter.

As the sun popped its yellow head up above the tree line the clay pot I had strung up all those months ago would burst to life with the zinging of worker bees embarking on a day of foraging. I would stand silent and still just behind it (never stand in front of the entrance to a beehive - it flusters them) and just observe, too afraid to try anything further on my own.

Before long I realized that the time had come to man up and open up the hive and for that I would need equipment. My first bee suit marked a rite of passage and with my beautiful mum in tow we set off into the urban wilds of Nairobi to purchase one. We ended up on the doorstep of the National Government Beekeeping Station and were duly introduced to a number of beaming officials (I have yet to meet a grumpy beekeeper). Through a man called Festus I purchased a fine white bee suit, a pair of thick leather gloves, a smoker and a hive tool along with the offer to join their small research team every Tuesday afternoon when they went out into the neighboring forest to poke about the twenty-odd hives they kept for "investigative purposes."

At last I was able to peer into the mysterious depths of my hive that had fortunately been colonized by a very pleasant swarm of bees and follow up with the few folk who requested my help. After scribbling down their addresses and a short brief on what they were looking for, I would stuff a bag full of sawdust chippings (fuel for the smoker) and head over in the evening (the best time to work with African bees) to offer my 'professional' insight.

In those early days I felt confident in my baseline knowledge but much less so when it came to the action bit. Most of the time I made things up as I went along and my clumsy movements often meant that the bees suffered more than was strictly necessary. But with each day I learned something new and before long I became adept at keeping my smoker alight throughout a session and swiftly learned to read the character of a swarm before opening up the hive simply by the sound and movements of the bees that zoomed in and out of the entrance. On several occasions I was even able to locate the reclusive queen scurrying about her brood comb and at such moments would giggle into my veil and utter noises that one would normally associate with the mentally unsound.

After several of these sessions I also discovered that one of the most pleasurable parts of doing this job was opening that door into this glorious 'other' world for the people who had requested my services. Before I lit my smoker and began with the bees - usually over a cup of tea and a biscuit - I would attempt to summarize the life of a bee with much gesticulating and bzzzzz-ing noises and then sit back and observe as stressed-out mums and overworked office dads transformed into wide-eyed children - just as I had done. I knew when I had nailed it because suddenly they would sigh and let their shoulders slump. With a faint smile on their lips they might then recline back into their chair, eyes raised heavenwards and utter a "phwoaarrr" or "wowwww" whilst the more articulate would exhale a soft "Incredible."

\*\*\*

My old friend the 'Bee Man' also received multiple visits from his self-appointed apprentice and by the end of the first month back home I had nine occupied hives of various shapes and sizes scattered about the bottom of the garden. Peter's right hand man was an elder Kikuyu called Bernard who sported a fuzzy mop of gray hair, a delightful smile and a calm, patient manner that I exploited to the full with a non-stop barrage of irritating questions. Thanks to Bernard and a daily exposure to the practical side of beekeeping I soon grew increasingly confident and began to spend more time alone in my new apiary.

Often I would stumble back long after the sun had sunk beneath the horizon with my red-beamed head torch attached to my veil and regale Mum and Dad with exciting stories of what I had witnessed that day. There was always something new to learn and after the shadowy days in Cape Town the solitude and peace of my surroundings worked like a tube of cool cream on sunburned shoulders. I still needed time to put myself back together again and to clear out the murky patches in my brain that still harbored hints of negativity and self-doubt.

The more time I spent with the bees the more insight I gathered into how they managed it; how some 80,000 individuals were able work together in a dark, confined space in such harmony. After musing over this for several weeks I learned it boiled down to a number of things primarily involving the bees innate desire to maintain unity as a singular 'whole', the strict division of labor and a range of mind-boggling methods of communication that included the transfer of chemicals through the sharing of food, antennae twiddling, noise, scent and dance.

It was this latter method of communication that 'hooked' most people and although the honeybee is well-known for performing a number of different routines it was the famous Waggle Dance that received the greatest number of votes. Also known as the Figure-of-Eight Dance this particular set of movements is performed by the female worker bee when she wants to alert her sisters to a particular source of food located more than one hundred meters away from the hive. To communicate this information the dancing bee will run ahead for a short distance on the vertical comb surface before turning around in a semi-circular direction back to the place she started out from. Waggling her abdomen along the way she will then turn again, completing the second half of the circle in a rough figure of eight movement that conveys to those watching the precise distance and direction of the food source in relation to the sun.

Amazing.

During the evening hours I liked to philosophize over how far we humans have strayed from this path of collective betterment and blended 'oneness' with our surrounding environment. When did we become so lost? What happened and why?

*How do we go back to the way things were at the beginning?*

The answer, or part of it anyway I decided, lies in copying the bees. We need to collaborate as individuals who share the same general characteristics and desire for the future irrespective of minor subtleties like race, class, gender or sex. We need to work out a system for exchanging information and knowledge and we need to unite, sharing the burden of hard graft and sacrifice in order to reconnect with the natural world that lives and breathes alongside us and without which, none of us stands a chance.

Even at the time I knew my theories on human deliverance were far too green to carry any weight in the face of 21st century reality, but nevertheless I remained undeterred for the simple reason that just three hundred meters from where I sat each night were living, breathing examples of creatures - bugs no less - who went about their day to day lives doing exactly this; delivering life, enhancing the world about them and proving that no matter how much time an individual is granted in this particular world, it is what is done during that stint that matters most.

* * *

Several weeks after arriving back home I was invited out to dinner with a mixed group of friends. As we set about demolishing large platters of chicken curry I found myself recounting a story from earlier that afternoon, half of which I had spent floating in the dam after my smoker had gone out and the bees had gotten disorderly. I survived with a swollen face and a deflated ego but it was well-deserved for I had begun to get cocky. The bees were not my 'friends', they were wild animals that had done well to tolerate my continual presence but that day I had pushed it too hard.

Lesson learned.

One of the ladies seated opposite me was the sister of a friend I had grown up with. Her name was Samantha, or Sam and for the past few years she had worked with a small NGO in the South Rift Valley region of Kenya.

Sam had started out in this particular area as a university student eager to set up a basic ecological monitoring program in partnership with the local Maasai community and several years down the line she was in charge of a fantastic grassroots organisation known as SORALO (South Rift Association of Land Owners) that had diversified to include ongoing monitoring of livestock, wildlife, indigenous vegetation and water supply in order to help the local communities identify changes in their environment and manage these key resources.

Over the years her persistence and energy had caught the attention of several international donors and together with a small team of local enthusiasts, the Lale'enok Resource Center was constructed - a humble collection of pale yellow buildings with thatched roofs, several canvas safari tents and a wonderful open air shower where one could scrub off the day's accumulated sweat under a glittering night sky.

Unlike the majority of NGO's I had observed working throughout the country, SORALO was run by a team of passionate young people (small salaries and smaller egos) who were on the ground 24/7, faced with a dazzling variety of situations that often called for a strong head and a compassionate heart over the requirements of multiple glossy certificates. On more than one occasion I had heard tales of meetings disrupted due to foreign tribesman toting poisoned spears who had crossed the open plains at night and made off with a herd of local cattle or a spate of crop-raiding elephants who in times of drought had ripped apart several shamba's, occasionally trampling someone that had gotten in the way.

The Lale'enok team at large recognized the incredible uniqueness of this area and her people and were constantly looking at new ways of working alongside the local people and for viable solutions to the issues that constantly feature in so many of Africa's rapidly dwindling wilderness regions.

Kenya's population is growing at a formidable rate and the spread of information through the mighty mobile telephone has impacted the community as much as the mass importation of *BodaBoda* motorbikes that now ferry multiple passengers across distances that once took several days to walk.

**Development is an unstoppable force.**

While the area now boasts several primary schools and a number of local businesses doing a roaring trade in imported Chinese clothing there has also been a sharp incline in incidents of charcoaling, poaching, land grabbing and of a variety of particularly corrosive human-wildlife conflict areas in a region where every living creature is dependent on the same narrow resource base.

As a family we had spent countless weekends in the area camping under the stars and bathing in the foul smelling waters of the hot springs that bubble up from the grumbling bowls of Mother Earth, exploring the primordial forests on the Nguramen escarpment and tubing down the sludgy brown waters of the Ewaso Nyrio River.

At the end of the evening Sam handed me her business card and mentioned that if I was interested she knew of a possible opportunity for me to put my recently acquired bee skills into practice with a women's beekeeping project that had been initiated several months ago by another NGO of less rigid vision. As far as she could recall the project had fallen flat after an initial delivery of 400 Langstroth beehives and if I fancied going down to take a look she'd be willing to make a few phone calls.

Bingo.

\* \* \*

Several days later I found myself bouncing along in a clapped-out Land Rover, heading south towards Lale'enok on a "rekkie" trip. Sam had given me permission to base myself at the center for a few days in order to have a sniff around at what was left of the bee project and if I liked what I saw she would be willing to back any ideas I had for how it could be developed.

Full of excited anticipation and a craving any opportunity to test my abilities in the field as the experts called it, I wound down my window and breathed in the cool, moist air that hung in a shimmering blanket around the base of the Ngong hills. Silently thanking the universe for this spectacular opportunity I garbled away to Davis, Sam's trusted driver as we whizzed down the escarpment road down into one of the most magnificent expanses of untamed Africa.

To get to Lale'enok one must travel 130km South of Nairobi on one of the crappiest roads the country has to offer. Leaving the city lights far behind, one does their best to dodge the potholes and overloaded *BodaBodas* that weave about underneath towering sacks of charcoal. Then one must be on guard for herds of lop-eared donkeys that saunter at whim out of the thick scrub lining either side of the road and have one hand ready to punch the hooter at the sight of a dik-dik, standing wide-eyed and trembling in between the holes as it processes the sight of the metallic beast thundering towards it.

A few bone-jarring hours later one then enters the Magadi township that resembles a Hollywood movie set worthy of Peter Jackson himself. The town itself, dominated by a large, imposing factory, consists of a dozen railway tracks, three belching chimneys, a sprawling nest of twisted steel scaffolding and a slowly rotating tube the size of a house. Attended by an army of workers resembling pale blue Oompa Loompas in matching overalls and hard hats, the air is thick with heavy white dust and the pungent tang of sulphur hangs heavy leaving you suddenly less eager for the boiled egg sandwich carefully set aside for lunch.

The Magadi township lies on the Eastern shore of Lake Magadi, a 100km square saline pan whose waters lie purple and sludgy beneath a thick crust of sodium carbonate, which precipitates a mineral called trona. This is sucked off the lake's surface in huge quantities and shunted through the trembling guts of the factory before emerging out the other end as soda ash; a vital ingredient used in the manufacture of glass, soaps and assorted chemicals.

As a result of the factory and its unremitting timetable, a bustling settlement has grown up in the surrounding area that boasts several orangey-brown apartment blocks, a number of ramshackle *dukas* that sell anything from mobile airtime to long expired packets of chewing gum and perhaps more surprisingly given the area's inhabitation by lion-killing warriors; a bright green public swimming pool.

After a brief stop for fuel and the last cold Coke for several hundred kilometers, Davis and I cautiously made our way over a twisted, bumpy causeway that dissected the lake and connected the town to thousands of kilometers of wilderness that stretched out in every direction along the floor of the Great Rift Valley. One could spend an entire lifetime discovering the spectacles on show in this region, including several alkaline lakes, bubbling hot springs and a batch of prehistoric volcanoes around which some of our earliest Homo Erectus ancestors once hunted.

Approaching the end of the causeway I stared in awestruck wonder at the sight of a billion flamingoes delicately poking about in the shallow lake waters, heads down sifting through the foul smelling sludge in search of the algae that lends their feathers that famous pink flush. Driving through their midst slowly I snapped away happily with my camera enjoying the feeling of hot air on my neck until the closest group of birds took fright at the sudden squealing of wet brakes. Suddenly they were on the move, churning the water with spindly legs until finally they achieved lift off... up, up and away into a sky so blue it hurt your eyes.

After crossing the causeway Davis engaged 4WD with a quick flick of an experienced wrist and we ploughed on through a vast landscape of open savannah that lay motionless in the midday heat. The grass at that time of year is long and golden in color and the road was covered in a six-inch blanket of pale, powdery dust so fine that one eventually emerges from the car the same color as the background vegetation. Scrubby acacia trees dotted the horizon at random, offering reassuring patches of dark green and beneath which we spotted several pairs of gerenuk -the long necked gazelle- standing patiently in the sparse shade hanging out for the evening cool off.

This area has long been a secret spot for the more reckless city dwellers craving a quick escape into the bush for an opportunity to test their suspension and fiddle with their Leatherman's. But it is the pastoralist Maasai people who inhabit the region on a permanent basis living in much the same way as their ancestors had done in seasonal huts made from mud and cattle dung. These *bomas* are surrounded by thorny fences cut from the branches of the vicious wait-a-bit trees that abound in the area and which protect their precious livelihood - cattle, goat and sheep during the dangerous hours of darkness. Wrapped in brilliant red *shukkahs* and clutching sharp, glinting spears these magnificent people roam far across the land with their animals in search of water and forage at the complete mercy of the natural world.

The Olkirimatian community consists of a flyblown, ramshackle marketplace, two dusty *dukas* and a primary school. It is the closest call to civilization for the Lale'enok Research Center that itself was a further 2km down the rocky track on the banks of the toffee-colored waters of the Ewaso Nyiro River.

Arriving slightly cramped and already covered in a thin layer of sweat/dust/sun cream, one can't help but feel the urban nerves relax and un-frizzle as the sound of the ignition is at last turned off. Once the engine had finished its popping and fizzing my ears were suddenly caressed by the sounds of the bush that descended from every direction; weaverbirds gossiping from their nests high up in the acacia trees, goat bells tinkling and donkeys braying in the distance whilst troops of baboons barked from the fig trees that lined the riverbank. As a trickle of sweat worked its way down my cleavage there was little to do but pull in a deep, hot breath and raise an appreciative eyebrow at the fact that places like this still exist in the world.

After a welcome cup of sugary tea I threw my bags inside an empty safari tent and wondered off to locate a man called Steve whom Sam had arranged for me to meet upon arrival. Steve was a striking Maasai, six foot tall with a narrow pointed head and a smile that split his face exactly in half. I eventually found him chairing a meeting with several village elders in the adjoining community center and once he was finished we exchanged the customary "Supai alengs." Dressed in the traditional red blanket tied around his waist with a fine beaded belt and a hunting knife Steve explained that he had worked at Lale'enok since it's inception and over the next few days it would be my honor to have him accompany me as my designated driver/translator/assistant as we crashed about the savannah visiting the bees.

Eager to get started straight away I topped up my water bottle from a tank filled with rainwater and mosquito larvae and jumped into the front seat of a clapped out Daihatsu jeep that appeared to be held together by a couple of sun bleached zip ties. Once back on the move our going was slow, mostly due to the fact that 'roads' as such didn't really exist but Steve, hunched over the steering wheel so as not to bump his head on the roof, was a real fundi at navigating us along in a direction that he was clearly familiar with – skillfully turning left and right around certain bushes or anthills that marked our passage like organic road signs.

As we crashed across sandy *dongas* and over gently rolling plains dotted with pockmarked volcanic rocks, Steve filled me in on his history at Lale'enok before explaining that he knew all about bees in the area for it turned out that he too kept a few traditional log hives back at his *manyatta*. Only last week he recalled shimmying up the trunk of his bee tree on a night with no moon "because the bees can see you when there is light" and without a suit or any of my fancy tools had managed to remove several kilos of honeycomb "with only one or a few stings."

I was dutifully impressed, having long heard of the skill of many of these traditional honey hunters found all over Africa and for several minutes barraged him with questions on the taste and color of the honey as well as the behavior of the bees.

Steve chuckled freely at the crazy mazungu beside him and when we passed a brilliant green bird sitting atop a nearby bush, he slammed on the brakes and wildly prodded my arm as the creature turned its exquisite head, eyeballing us haughtily.

Steve then poked his head out the window and uttered a "chirrup chirrup" sound that caused the bird to cock its head. After a few more noises the bird puffed out its chest and sang back "chirrup chirrup." Beaming widely Steve then turned to me and explained that this was the famous 'honey guide' that would often follow the herders as they walked the plains with their cattle, hopping from bush to bush singing for attention. If the herder in question knows what they were doing, Steve explained, they would answer the bird and leaving their flock in the hands of someone else they would follow it to a place where the bird had located a wild beehive. The human would then set about harvesting the honeycomb that would guarantee him a loving back massage from his wife that evening along with several days worth of homemade honey beer that would be set aside for the next 'special' occasion.

Pressing the accelerator to the floor once again Steve then gave me an overview of what little he knew of the failed bee project confirming that several months ago a large quantity of expensive modern beehives had appeared on the back of a truck and were duly offloaded by a group of sweating people from the city. The hives were a donation from a well-meaning NGO (no one knew which one) and were duly assigned to the Olkirimatian women's community as part of a project that would have inevitably included the words 'EMPOWERMENT' and 'GRASSROOTS ENGAGEMENT' in its proposal. However, after the initial distribution of the hives there had been no further communication, training or additional equipment delivery and after several weeks the beehives that had managed to survive the hungry jaws of termites were handed out at random and were now strung up in various trees in an area the size of London.

Steve knew of several women close by who had received a few of the hives in question and our mission that day was to locate them and establish what the situation was: How had the hives been placed? Had they been occupied? Did the women know what to do next? What was the plan man?

Just after 2pm we finally arrived at our first *boma* and were introduced to a large group of chattering women dressed in brightly coloured *kangas*, each with enough beaded adornments to decorate a Harrods Christmas tree. After a brief round of introductions Steve and I were then led a few hundred meters away to a dense thicket of thorny bushes where the beehives had been placed. After several minutes of cursing at the prickly foliage we soon found ourselves staring up at a dead tree from whose blackened branches hung two rather battered-looking beehives. Impossible to access due to the fencing wire that held them firmly shut several meters off the ground I failed to spot any sign of movement at the entrance and turning to Steve we exchanged disappointed shrugs.

Returning to the *manyatta*, Steve and I were then invited inside to share a cup of goat milk tea and within seconds two wooden stools had been bought forward by a pair of giggling children and set beneath the shade of a scrubby acacia. We were then handed two enormous mugs of steaming, milky goodness that I promptly discovered contained an entire sackful of sugar. With every tentative sip I felt my jaw thud with the approval of 100 billion bacteria that would no doubt spend the rest of the day burrowing through my enamel.

Eventually I managed to drain the contents of my cup, my head buzzing with the heat and the musical notes of rapid fire Maasai and forty-five minutes later we extracted ourselves from the group of beaming women. Waving fond farewells to the kids who had placed their favorite baby goats one by one in my lap, we set off in the rattling Daihatsu to our next destination.

That afternoon we visited two other sites similar to the first, each time wading through long introductions and then inevitably stumbling about in a thorny thicket checking out the uniformly empty beehives. By 5pm I was sunburnt, exhausted and suffering from what I feared were the first pangs of adult onset diabetes when suddenly I heard a shout from Steve and went scurrying over to find him peering up at a hive that hung from the low lying branches of a young fig tree. Moving in closer I picked up on the telltale movements of worker bees zipping in and out the entrance and we high-fived one another in delight – they do exist!

Desperate to put my bee suit on and prove to Steve that I wasn't just another stupid tourist I lit my smoker in 2 minutes flat and scrambled up the tree to carefully unwind the wire that held the beehive in place. Removing the lid I peeked inside and was greeted by the sight of a medium sized colony of bees happily going about their business. As they caught on to the fact that something peculiar was at play I watched several bees raise their shiny abdomens into the air and listened to the increased pitch of their buzzing.

Robert had taught us that this movement signaled the bees releasing a cloud of alarm pheromones into the air that would slowly spread amongst the colony as a warning signal. The longer the hive remained open, the edgier the bees would become and as we had a while to go before the sun went down and the bees settled for the night, I decided to close the lid and leave them in peace.

Back at camp, flushed with success I bid Steve goodnight before joining the rest of the camp for a fine dinner of Nyayo beans and posho before the highly anticipated shower. Feeling instantly refreshed I then made my way to my tent by the light of a head torch and lying naked on my mattress slowly drifted off to sleep accompanied by the scuffling, bellowing, roaring sounds of the nighttime wild.

*

The next day dawned bright and by 6:30am I was covered in the first layer of that day's sweat. After a quick breakfast of overripe mango and ginger biscuits I met Steve and off we sped to another distant corner of the South Rift Valley.

The day progressed much like the one before: lengthy introductions, multiple cups of tea and the exploration of impossible–to–access hives. None of the women knew what next to do with their hives and as we slowly made our way back to camp later that evening I couldn't help but feel daunted by the task we faced. Everything about this place (harsh, hot and desolate) seemed totally incompatible with the image of beekeeping I had in my head. In one afternoon alone we had come across three colonies that had quite literally 'cooked' to death and the majority of the hives that I had been able to open and inspect contained an assortment of creatures more adept at surviving life in this semi-desert environment including three fat scorpions, a pair of electric green lizards and a family of mice with long noses and bushy tails that had squeaked in indignation as I pulled apart their nest.

The distance between the women's *bomas* and the terrain we had to cross in order to get there was equally problematic. It took us the better part of two days and four tanks of fuel (courtesy of a leaking tank) to visit just a handful of the women who had been given the hives and apart from Steve, I was the only person in the area with the faintest idea of how to move forward.

At that point I couldn't help but feel a surge of irritation at the fact that in taking up this project I would in essence be clearing up someone else's mess. What if I failed? Would I then be labeled as yet another Great White Hope, full of big plans to EMPOWER and ENGAGE? One who would inevitable run out of time, energy and motivation as bigger and better options popped up in the more civilized world? But despite any concerns over my own abilities there was a faint tingle for adventure. Here was a chance to play and learn, an opportunity to explore and test my skills in an environment that made me feel more alive with every second that I breathed in that warm, dry air. There is always a way to improve a situation I whispered to myself as we bounced along another dusty track and with what Oprah Winfrey might define as an 'Ah Ha Moment' I came to the sudden realization that if indeed I was serious about my theory that young people might take over the world and make it all better again I had better get used to the reality that much of the work that lay ahead would inevitably involve the sorting out of multiple cock-ups instigated long before we had even appeared on the scene.

Thus, upon returning to Nairobi I wrote out a six-month project proposal entitled 'Bees for Conservation' and sent it off with a flurry of emails to Sam. Yes, I was keen. Yes, I would give it a bash and yes, I was happy to work for a minimum salary. When can I start?

Several days later I found myself back on the crappiest-road-in-the-world as the official Project Manager for the Bees for Conservation project. The aim of this six-month trial was to test the viability of the area for a community beekeeping project and to arm the women with as much information as I could on modern beekeeping methods.

Sam had put the word out through her network of people on the ground and the afternoon following my arrival there was an official introduction and welcoming ceremony laid out in the community center complete with free sodas and freshly baked *mandazis*, reason enough for the entire village to appear dressed in their finest collection of cloth and beads. Walking into the community center I was swiftly escorted through the masses of chattering women and encouraged to take a seat on the only plastic chair in the room.

I was introduced formally to the village chief, the chairlady of the Olkirimatian women's group and several prominent village elders who all lacked teeth and whose ebony black skins were creased like well-worn leather loosely arranged over pointed elbows and knock-knees. As I settled into position the group bustled about in a dazzling explosion of patterned material and the harmonious tinkle of multicolored arm cuffs, anklets, necklaces and earrings and as they approached one by one I did as Steve instructed; lowering my gaze and allowing my forehead to be 'blessed' with the briefest of touches.

And then the speeches began.

An hour into the occasion, during which the sun had climbed to its zenith and was busy pounding on the metal roof above our heads I was struggling. My Julia Roberts smile that had done well to stay put for the first twenty minutes had morphed into a lopsided grimace and the sea of quizzical faces staring up at me from the cement floor swam before me as I fought to stay awake. Sweating, my thighs had stuck to the plastic seat and I was dying for a glass of water but as most of the women had walked over three hours to be there that day and no one was eating or drinking anything I felt too embarrassed to ask. When the chief finally did sit down to prolonged applause the chairlady then stood up and took over. Oh God.

Two long hours later I was as close to death as ever before when a sudden rumble of conversation rippled through the room and two hundred women suddenly jumped up and burst into song. Accompanied by much clapping and high pitched ululation I found myself clasped to the moist bosoms of several of the ladies who encouraged me to join in the jumping, gyrating column that exited the center. With a final whistle and clap I was officially applauded as the latest savior of the 'International Olkirimatian Women's Official Bee Committee Delegation' and was handed a small blue beaded necklace as a welcome symbol of friendship and good luck.

Unfortunately it didn't end there. My dreams of a nice cup of tea and a soothing swim in the river were postponed as we, the official delegation, were ushered into a neighboring tin shack (inside temperature 300 degrees) and served large aluminum bowls piled high with mounds of glistening goat fat accompanied by a plastic jug filled to the brim with Bilharzia-riddled river water.

And so the project began.

Funding was our immediate constraint but as we already had the beehives in situ it was possible to get by in the earliest stages on the bare minimum. We started off by translocating several of the abandoned beehives into one central apiary located 100 meters away from the main Lale'enok buildings and which consisted of nothing more than a few fence posts sawn in half and dug firmly into the ground. Around the base of each post we then nailed a metallic skirt to deter the marauding honey badger along with a few empty tin cans topped up with river water and twigs, which served the purpose of 'emergency ladders', should one of our striped mates loose their footing whilst taking a drink. Finishing off we gave each hive a good smearing of beeswax mixed with a fistful of wild basil that would act as an attractant and sighing deeply with satisfaction we returned to camp in time for yet another cup of tea.

Over the next few weeks Steve and I set up several similar apiaries all over the area in a number of locations suggested to him by the members of the bee delegation. Word got around pretty quickly that a mad mazungu was on the loose and I was promptly nicknamed 'Mama Nyuki' - the 'mother of the bees' a title that stroked my ego more than anything else. Any time we stopped for a packet of gum or a soda in-between apiary-building missions we would be approached by interested individuals who would come to shake my hand and let their children touch my hair, skin and clothes with bashful giggles.

Once three apiaries had been completed in this way I returned to Nairobi, making back to Lale'enok every few weeks to check on their progress. During this time Sam managed to raise a small amount of supplementary capital for the project and after a brief meeting one evening we decided that the best way to utilize these funds would be to host a three-day beginner's beekeeping training. Our aim of the 'bee festival' as it came to be known was to give the women as much information as possible in the hopes that they would then claim responsibility for the project and be able to move ahead without the need for long term interference.

After much back and forth via Sam, Steve, the Chairlady and several members of the official committee, the dates for the festival were fixed and my attention shifted to figuring out what information would be necessary for the group and how to get it across. The women were largely illiterate and spoke no English or Swahili so my new hero Steve would be required to play the role of translator, explainer and general everything man.

The day of the festival loomed closer and back in Nairobi, in the cool confines of Mum's kitchen, I took a leaf out of Deborah's books and set about making a range of honey flavored delights for the women. I then packed Dad's car with ten complete beekeeping outfits purchased from a local shop down the road and topped up a number of jerry cans with fresh rainwater from our tanks.

\*

The festival kicked off as I envisaged: with several lengthy welcoming speeches followed by much clapping and singing. Using the shuffling of their feet, the clapping of their hands and the jingling noise of a million beads to accompany their beautiful voices, the women wound their way about the center in a sensual line of flashing smiles and swaying, multi-colored hips. Twenty-eight women had arrived at the center on that first morning, some having walked overnight from distant homesteads on the far side of the conservancy area. Each arrived dressed in their finest outfits once again; brilliant red, orange and blue fabrics draped over slender bodies in a variety of styles. Several of the women had one extra cloth slung across their shoulders within which nestled a tiny baby. Never before had I witnessed such well-behaved children and throughout the three days those tiny, shiny humans would lie peacefully strapped across their mothers' chests occasionally waving a fat fist in the air or mewling for milk.

It took the women several minutes to settle into place as they danced around chattering like exotic birds, allowing me time to absorb the fact that I was standing in the vicinity of a group of people so far removed from my western ideologies that we might have belonged to a different species altogether.

Clasped on each foot above a pair of uniform plastic shoes were six-inch beaded cuffs that matched the ones on their forearms, necks and wrists. The older members of the group, characterized by their purple, rheumy eyes and wrinkled faces, sported closely shaved heads with stretched ear lobes grazing their shoulders, but regardless of age each woman also carried a small leather pouch around her neck inside which lay a mobile phone that would beep and flash every couple of minutes. No amount of persuasion from neither Steve nor or I could get the ladies to switch their phones onto silent mode and throughout the festival, the incessant Nokia ringtone proved the greatest distraction.

With much clattering of chairs and clearing of throats, the festival began with a brief background on the three castes of honeybee, before moving on to the kind of products that can be extracted from the beehive. Several members of the audience had already fallen asleep by that stage but as soon as I passed around some beeswax candles, body cream and a tub of propolis ointment the ladies snapped awake and excited chatter filled the room. Some time later, after a cup of goat milk tea sweetened with honey and an accompanying flapjack, I discovered a sort of flow and with Steve at my side translating every word, we soon figured out a way to keep the ladies engaged.

After a late lunch of the infamous goat stew and a heat-dazed siesta, the group were divided into three and the first group of ten was then ushered off to change into bee suits. Once zipped up the women looked hilarious, instantly transformed from visions of timeless elegance into a pack of saggy white-suited zombies. Once the rest of the group had stopped shrieking in delight at the sight of their comrades, our small band traipsed down to Apiary 1 and while Steve and I bobbed about in the background practiced lighting the smoker before easing off the lid of one of the occupied hives. Ten sets of *ooh's* and *ahhh's* filled the late afternoon air and my heart bubbled over with pride and, up until a mobile phone rang and several bees soared upwards like miniature missiles ready to challenge the noise, I figured things had gone pretty smoothly.

At the end of the practical session our group returned to the community hall and as I set about packing up my notes for the day, eager for a swim in the river and a few moments of peace and quiet, I was approached by the Chairlady, a wizened creature with dark twinkly eyes and stretched earlobes that swung lazily to and fro at the slightest movement of her head. Pulling me aside I managed to work out that the women had gotten chatting over their lunch break and had decided that a daily monetary bonus was required to recompense them for attending the training. The Chairlady looked bashful but was firm in her expectations and wanted to know when they would be receiving the money and how much they should expect.

Incredulous, I stared at her. What? The women were getting free accommodation, free food and drink, free suits and free reign on a project that they could then take as far as they wanted. Why on earth would we pay them to do that?

And then it clicked. I had been naïve to think that so many women had the spare time and desire to spend three whole days away from their homesteads listening to some white kid from the city talk about bees. It was a crushing blow to my ego and upon informing the Chairlady that there wasn't any extra money for stipends I retreated to the central camp area to loud rumbles of dissent.

During a panicked phone call to Sam that evening, she apologized for not seeing this one coming, informing me that several other 'do-gooder' projects in the area had started paying local people to attend seminars and clinics as a way of using up their budget. We were both embarrassed and saddened that this mentality had spread to 'our' women, but at that stage there was nothing left to do but stand firm and carry on.

The following day and with some trepidation I walked up to the community hall and discovered that only half of the women remained. Left with little choice but to continue I carried on going over the pests and diseases one had to look out for in the hive before a beeswax candle–making experiment that coaxed a few smiles out from behind clouded expressions.

That afternoon, Steve and I took the remaining women back down to the apiary and once again went through the motions of opening up the beehive. There's nothing like bearing witness to a few moments of shared connection to lift ones spirits but I was tired and by that stage was looking forward to the whole thing being over.

Back in Nairobi I had downloaded several bee movies and clips off YouTube and later that evening, with a handful of women who had agreed to stay on at the camp we set up the projector screen and watched a series of short bee films from around the world. As the majority of films were silent our soundtrack was locally provided by a chorus of screeching bush babies, the distant 'whoo-oop' of a hyena and a barrage of giant black moths who cannonballed themselves into the screen with single-minded intensity.

The final morning was more of a "thank you" and wrap up that even fewer women attended. We handed out the bee suits to designated individuals from each area and divided up the bright yellow wax candles that had set perfectly overnight. After a few shorter speeches the festival was officially closed and the women were packed into the back of an open topped lorry to be dropped off at their specific village centers. Suddenly alone in the dusty car park and surrounded by silence once again, I couldn't help but feel relieved.

I had to be back in Nairobi that evening so after clearing up the various posters and chairs that littered the center and thanking Steve profusely for his continued hard work and patience I repacked the car and set off back north. Alone with my thoughts and with the window wound down to harvest the faintest of breezes it didn't take long for me to lose my frown and sit up straighter, staring out at the shimmering land I had been so fortunate to experience over the past few months.

With a final deep breath I put the stipend fiasco aside, plugged in a fat reggae soundtrack and sang my way back up the escarpment.

**Next time I'll do better.**

Returning to the mother city, my body was dirty but my mind sparkly clean. An overwhelming appreciation of the world and the beauty it contained filled my insides and although I knew how far the project had to go if I really wanted to turn into something more than just a 'nice try,' I felt satisfied… and even just a little bit proud.

**Everyone's gotta start somewhere.**

# Today is about reality

# CHAPTER TWELVE

A few weeks after the bee fetival I jumped back in the car with Davis who was heading down to Lale'enok on a resupply mission.

Eager to escape the city, I was looking forward to checking up on the bees and discussing with Steve a number of options for how we could solve a recent crisis that had taken its toll on Apiary 1. He had called me one evening and told me that a resident troop of baboons, the same ones that had begun to get cocky and enter the camp, bouncing on tents and stealing from the pantry had also discovered our beautiful beehives. They had started off by hurling wild figs and river pebbles at the strange white boxes, slowly working out which ones contained the creatures that made the strange buzzing noises and stung if anyone got too close. They had then progressed to messing around with the empty ones, knocking them to the ground and ripping apart the spindly wooden frames that lay inside.

Back at home I had scavenged a number of old bicycle inner tubes from the dusty corner of Dad's store - Steve's idea for securing the hives in position - and armed with my trusty bee book (there was still so much to learn) Davis and I roared off down the escarpment beneath a wide open sky the color of cornflowers.

Upon arriving stiff and hot in the Lale'enok car park I was too preoccupied with gathering the tubes into a bundle and finding my pocketknife that had slipped under the seat to notice the shiny black Land Rover with tinted windows that was parked up beside us.

Waltzing into camp I greeted Loimpia the camp cook - a man blessed with the ability to make tinned beans and bendy butternut into a Master Chef classic - and several of the young research assistants who were sat behind their computers downloading camera trap photos. The office felt unusually quiet. Over dinner that night as the Lale'enok team sat around the camp table over a bowl of 'Vegetable Glory' (Wednesday was market day when you could buy all sorts of lovely vegetables but by Friday they had all gone spongy) we were joined by a stranger who pulled up a spare chair and helped himself to a bowlful. Ignorant about who this strange man could be and what he was up to (kind of good looking in a polished sort of way) I chatted away merrily garbling nonsense and recounting the baboon/bee story to anyone who cared to listen.

The man (I forgot his name, let's call him Tarquin) wore a clean pair of dark blue jeans and a white V-neck shirt that oozed Calvin Klein vintage. He had pushed back a pair of gold rimmed aviators atop a mop of dark hair flecked with gray and sported a five-day old 'man' beard that I found rather attractive.

On discovering that he was indeed the owner of the racy Land Rover outside I figured that he might well become my husband and decided not to describe to him what it felt like to be stung between the eyes (guaranteed Neanderthal brow for four days) and did my best to spoon stew into my body as opposed to over it.

But the atmosphere around the table was weird. My friends, who would ordinarily be engaged in a fierce debate about 21st century conservation initiatives or the best way to dart a lion, were instead sitting around quietly chewing on their food and occasionally murmuring the odd "pass the salt, please."

And then I discovered why.

The strange man who sat with us was a scout for a multinational oil and gas exploration team that had recently received permission to poke about the area and dig a few holes. At first I didn't really understand and carried on gaily asking questions: What did that involve? When, how many and how long for?

It was only later that evening when I lay in bed listening to a family of bush babies scream at a prowling genet cat outside my tent that the first tingles of fear made themselves known.

When I returned to Lale'enok some weeks later I was fully prepped by Sam: The oil and gas guys had officially moved in and set up an enormous camp a few metres away from the community center. As a result, they would be paying a daily 'conservation fee' that would be channeled into one of the women's projects - always a good thing I figured, but unfortunately the rest of the story wasn't.

That afternoon failing to find Steve, who was out running errands, I couldn't help but have a snoop around the posh new campsite. Their tents were big and shiny, the mess area covered an area of fifty square metres and was filled with long white tables and matching chairs. There was a team of six chefs busy in the well-equipped kitchen and the steady roar of a generator throbbed in the background. The grass inside the compound had been cropped short, wild scrubby bushes had been trimmed back to stumps and a vast pile of plastic water bottles lay in a heap in the furthest corner by the solar showers.

Around 6 o'clock a cloud of dust appeared on the horizon and one by one a steady stream of glinting 4X4's made their way into the camp disgorging several men in reflective jackets, dark blue jeans and crisp white shirts.

Suddenly I didn't want a husband.

That weekend marked the official end of my six-month pilot project. During that time I had proved that there was indeed potential for an extended beekeeping push in the area but it required a lot more funding and input to make the project 'viable.' I was eager to continue and had sent off a number of funding requests but nothing had come through yet. "Hang in there," Sam had told me.

On my last full evening at the camp having given Steve a large pot of home harvested honey to thank him for his help, I decided to take a stroll alongside the riverbank to reflect, breathe in the fresh evening air and watch the distant rain clouds gather on the horizon. Walking swiftly past the oil and gas camp I concentrated on the swallows that dipped and dived in the air above my head as I listened for the cicadas busy warming up their musical legs before their nighttime performance kicked off.

After several hundred meters I was passing the last of the white tents when I noticed several discarded beer cans that had been shoved unceremoniously down one of the openings of a beautiful orange anthill. Kneeling down I slowly wiggled them out with a branch, taking care not to destroy any more of this Fantasia-like tower and glancing up managed to strangle the urge to hurl the whole lot at a group of men holding clipboards sauntering past on their way to the mess tent. A strange red mist then appeared to settle over my vision and shaking with rage I waited until the men had disappeared from sight before bending down beneath the temporary wire fence that marked the boundary to the camp. Walking towards the closest tent I unzipped the flap and hurled all eight cans plus a few dried out wildebeest droppings on to the neatly turned bed.

But the fury in my throat failed to subside and on returning back to the track I picked up a twig and smacked it repeatedly against a rotten tree stump until my arm hurt and I felt better. All that exercise did wonders for the brain however and several deep breaths later I had a series of thoughts that at the time felt profound enough to warrant scribbling down.

**We humans are a product of our environment. As long as there are green spaces for children to play in, trees to rest under during the heat of the day, clean water to drink from and the buzzing, chirruping noises that accompany an intact ecosystem, life is good, we will be happy and**

our society will continue to plod ever so slowly forwards.

But if we continue to abuse our natural environment, using it to fulfill our own selfish needs and then tossing it aside once we are through before moving on and finding somewhere else to screw up, we will inevitably find ourselves in the same position as those beer cans: suffocated, bruised and face down in the dirt.

There is so much work to be done.

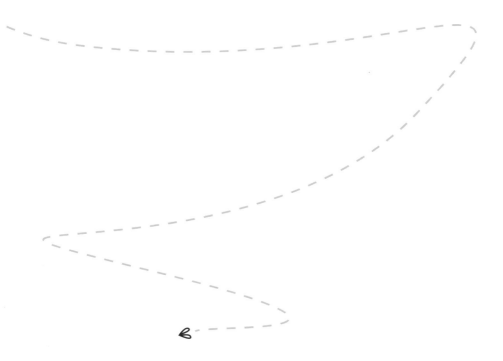

# Today is about Continual Growth

# CHAPTER THIRTEEN

Despite all life's little challenges, I knew I was onto something with the bees and as a result found myself relaxing into a steady routine, which included many miles on my morning run. Fresh air and exercise only added to my emerging confidence and soon I was signing up for an assortment of weekend competitions.

And so it was that I found myself one Saturday morning lined up in the shallow end of an algae-filled swimming pool at the University of Nairobi for my first ever triathlon. Ordinarily we would have dived in at the starters whistle but given that the water levels were just below knee height, the 'swim' section of the race would involve a ten-meter wade/dash before we slid down into the deep end.

The forty kilometer bike leg was no less hardcore given that our route directed us up and down a stretch of university avenue that heralded similar volumes of foot and vehicle traffic to rush hour in Mumbai and by the time I had completed the final ten kilometres run the build up of chafe between my heavily lathered thighs had reached chronic status. But nevertheless, given that the majority of my countrymen were not taught how to swim in their youth nor owned a bike designed for speed I had managed to beat all four of the other female competitors and had returned home stiff and sore but flushed with that particularly post exercise buzz that somehow justifies all the pain and makes you want to do it again.

Later that evening, my cheeks still aglow I received a phone call from a man called Chrispine (who unbeknownst to me had just become my

'Coach') who announced assuredly that I had just qualified for Kenya's first ever Triathlon team set to compete in the 2014 Glasgow Commonwealth Games in just under sixteen weeks time.

*Whaaaaat?*

## Glasgow

Waiting nervously in the starting tent I couldn't help but glance around at the other competitors intently focused on the next 3.5 hours of their lives; stretching impossibly toned legs and arms, sipping silently from branded juice bottles or receiving a last minute rub down from their coach. Staring down at my own body snugly contained in red and black Lycra I whispered out my name and country code that stuck out on my belly thanks to the enormous breakfast I had forced down earlier that morning

*I thought that was supposed to have digested by now?*

Irritated that somehow I had already managed to smudge one of my tattooed race numbers– number 30, I tugged at the swimming cap squeezing my temples and tried to breathe deeply.

The crowd in the grandstands located no more than twenty meters away from where we were corralled shouted and cheered and the morning air vibrated with electric anticipation. Some of the finest female triathletes in the world would shortly be stepping out onto the carpeted start ramp to battle one another in one of the most prestigious events on the race calendar.

Concentrating on the deeply defined muscles on the back of girl number 22 I swallowed hard, closed my eyes and forced in a few measured breaths, desperate to control the sudden surge of adrenaline before it got used up.

*How had I got myself into this?*

This was certainly not the first time I had asked myself that question but with less than fifteen minutes remaining before the start gun fired it was perhaps the most urgent. I was half expecting one of the neon-vested race officials to step up beside me, place a fluffy white towel over my shoulders and lead me out the back of the tent apologizing profusely for the dreadful error that had gotten me there. But as it turned out the only person who came close was one of the medics from the Australian team, who tapped me on the shoulder and with a look of bemusement in his eyes pointed out the empty energy bar wrapper stuck to my butt.

Lowering myself onto one of the thoughtfully placed chairs that lined the tent I helped myself to a chilled sports drink that guaranteed me my RDA of artificial electrolytes, synthetic sugars and brominated vegetable oil.

*Whaaat?*

Closing my eyes I put the drink aside and allowed my head to flop back onto my shoulders, focused on suppressing a sudden wave of nausea that threatened to disrupt the former Olympic gold medalist meditating in the corner.

Suddenly the DJ cut the background music forcing me back to the present and a murmur of conversation rippled through the crowd as the first woman - Andrea Hewitt - was called forward, followed by a lengthy description of her current achievements to date:

*"Five times Woooorld Champiooon!... 2012 Olympic Siiiilver Medalllllist!..."*

Glancing over her shoulders she eyed the rest of us with a gleam not dissimilar to a python about to give its victim a final squeeze and with a jaw set like concrete she jogged out of the tent, every square inch of her body rippling in time to her footfall. Up next was an English lass; tattoo number 7. Her red and white suit fitted perfectly around a body so lean and ready that I couldn't help but glance down at my own again and wish that I had trained just that little bit harder.

I was last in line and save for the official with a clipboard now alone in the tent but when the announcer did finally call my name followed by an enthusiastic *"Frooooom Kenyaaaaa,"* the crowd did their glorious whooping, cheering thing and I stepped out into the dazzling sunshine.

**There's no turning back now.**

Concentrating hard I attempted to lope down the starting ramp with the same confident light footedness I had watched in the others. The dark blue carpet beneath my bare feet felt luxurious, flags flapped in the breeze, strangers shouted encouragement and as I passed beneath the starting banner on my way to the lake I noticed my name flash onto the screen as a dozen TV cameras swung around to zoom onto my face.

*Where were mum and dad and Al - Could they see me?*

Twenty seconds later I had settled into my start box beside the other girls, my red and green tracksuit folded neatly in a plastic tub as the sun bounced off my shoulders. It was an unseasonably warm day for Scotland meaning that all those weeks training in a borrowed wetsuit had been in vain. Twisting around for one last scan of the crowd I inhaled deeply and told myself that I was going to be fine. Just fine.

On your marks…

Get set…

BANG.

Hurling myself into the water my goggles slipped a few millimeters on impact and my nose immediately filled with water.

*Keep going, it doesn't matter, go go go…*

**Five seconds into the race and although my vision was blurred I could already feel myself slipping behind, the furious wake of effervescing bubbles left by twenty-five pairs of feet gradually growing faint enough to see the waving green weeds that covered the bottom of the lake.**

*Focus. Rhythm. Breathe.*

Twenty seconds into the race and I was alone save for the thrashing limbs of a Mauritian girl just a few meters ahead. Lurching up from the lake's surface to make sure I was still on course I caught sight of the first marker buoy floating in the distance, its base distorted by the whirling arms of the lead group impossibly far ahead.

*Swim, woman, SWIM!*

By the time I reached the third marker buoy I had two hundred meters remaining of the first lap and had settled into a novice's perception of a rhythm; *inhale, stroke, exhale, stroke, desperate for breath, shallow inhale, stroke, GASP*. My heart burned but somehow my arms and legs had switched to automatic.

*Only 50 kilometers to go.*

*Oh my God, I can't do this.*

At that moment I couldn't stop my mind from flashing back to the mornings down at Lale'enok where I had forced myself to slip into the main current line of the *Ewaso Nyiro River* and swim furiously for forty-five minutes, only permitting myself a moments pause when one of the forty pound catfish that inhabit the river blundered into my chest causing me to inhale a lungful of brown, soupy water. Or when the monkeys hanging out in the top of the fig trees had begun to hurl enough of the soft, round fruit at my head that my heavy breathing pattern was disrupted. Then there were those mystical pre-dawn runs – before anyone else at camp had woken and when the savannah lay silent and cool and the fever trees floated above a band of morning mist like ghostly apparitions. At that time the only sounds aside from my pounding heart were the far distant whoops of the zebra and the invisible hoof beats of gazelle and giraffes.

*I had trained pretty hard.*

Or so I had thought up until the moment I first stepped into the athlete's village and been surrounded by men and women of all shapes, sizes and backgrounds: weightlifters, hockey players, sprinters and gymnasts; muscles, tan lines, team tracksuits and an eye watering selection of equipment - a fascinating insight into a life of pure dedication.

In the days before our race my three teammates and I had been instructed to do as little as possible other than to rest, stretch and prepare ourselves mentally, thus we chose to spend a few hours each morning and after-noon on the soft patch of grass located outside the main dining venue.

Here we would watch and listen, enviously monitoring the rippled mass of the Jamaican relay team who were digesting their high protein breakfasts close by (and who would later be joined by none other than Usain Bolt himself).

*Come on, Jess. CONCENTRATE.*

I was now half way through my second lap, confident that at least I was going to finish the swim section, but after another quick glance at the water ahead I failed to sight the Mauritian swimmer which immediately sent my brain whirring down a dangerously despondent path.

*I'm officially the last person in the race.*

*Tired... so tired.*

And then just like it does in the movies, a face swum up through the blackness urging me on and providing me the much-needed oomph to complete the final four hundred meters. The image belonged to a young African teenager that I had met some months ago back in Nairobi. His name was Charles and he lived in one section of the sprawling Dandora slums making his living from selling second hand clothes from a wheelbarrow. I had managed to get him and two of his friends up on stage to speak to a gathered audience about one particular story that had flagged my attention and which he duly did with flabbergasting success.

Inspired by the globally renowned TED talks and of course, the bees, I had set up a monthly event for young people with cool ideas for social change and thus the 'INSPIRED Talk series' was born. Each event hosted six – eight pre-selected speakers who came together for an evening of story sharing, networking and to spread some much-needed good energy. Kenya was experiencing a particularly rough patch at the time (thanks to some flustered jihadists) and although there had been a few teething issues at the start, the talks were slowly picking up momentum. The last event in particular had pulled in a large, vibrant crowd and it was perhaps the first time ever that I had found myself in a room full of people drinking in the words of a few young visionaries who were actively pursuing their own philosophies on this notion of a 'brighter future.'

It was Charles who had really nailed it.

With a beaming smile and slow but confident speech he had informed the audience of his current project in the heart of a neighborhood that he and his mates had hung out in since childhood. Bored of the miserable surroundings, the rotting garbage and the terrible odor he, with a few of his friends who shared his vision, had formed a group known simply as 'Mustard Seeds' and together they had applied for a small grant. Upon receiving the money they then spent every spare moment of free time upgrading one particularly large, filth-strewn area; clearing it of the unwanted stuff followed by some landscaping, a mural and a couple of benches where people could sit during the day. "We have changed the whole neighborhood – people now come here to rest and play here and even there is less crime than before." And then the words that in the middle of Scotland one hot summer's day had caused me to find that previously untapped sixth gear:

"For me and my guys, life is what you decide to put in – we only get one chance to do our best. Always be proud of what you can achieve with just a small piece of effort."

*YES, JESS. EXACTLY. GO. MOVE. NOW. HURRY*

Staggering out of the water back up the start ramp towards my waiting bicycle the crowd cheered - in fact they went wild. I tried to wipe the look of pain off my face and replace it with one of determination and when I saw that there was not one, but TWO bikes remaining in the transition zone my pace quickened yet again. Whatever glimmer of athletic professionalism I had managed to retain instantly evaporated as I punched a celebratory fist in the air and wheezed out a breathless, "YEEAHOOOO..."

The bike bit was by far the weakest part of my race and the first ten metres of spinning *whooosh* past the crowds caused an instant stiffening in the thighs that must have looked far funnier than it felt in reality.

The 8 km course – of which we were to do five laps, had been designed by a sadist and just as I was getting comfortable in the saddle the gradient pitched upwards for two hundred meters and my respiratory system went into meltdown.

*PEDAAAALLLLL!*

Completing that first lap was nothing short of miraculous for I was spent. I had lost my rhythm, my vision was blurred and I could feel my upper body rolling about on top of the bike like a poorly set cheesecake. Approaching the top of the first hill again I tasted bile in my mouth but far worse was the sudden roaring noise that erupted from the crowd I had passed just seconds before.

*The Mauritian!*

She passed me in the time it took to wipe off the ocean of sweat dribbling into my eyes. I tried to match her pace and stick to her back wheel maximizing her draft but to no avail, I couldn't hold on. With two hundred meters remaining of the second lap I knew with finality that I would be passing my family as the official back marker of the race.

*Nothing matters, Jess. Nothing matters.*

Fighting back tears I swooshed up onto the platform and off again in a blur of spinning pedals but just as I was preparing to shift my gears down into granny for the dreaded hill, a "whoop whoop" noise sounded loudly in my ear followed by a series of flashing yellow lights. Suddenly I found myself surrounded by men on motorbikes, which on any other day would have been most welcome indeed, but alas on this occasion confirmed that my race was over.

Feeling the first wave of emotion hit before my feet had even unclipped from my cleats I looked over my shoulder and watched the lead pack of girls hauling down the road towards me in a 45km/h mass of pumping muscle, flared nostrils and glinting carbon fiber. Passing with the same vacuumed suck as a fast moving truck I stood there mesmerized by the sheer power of the human body, watching them fly up the hill and around the corner as if they had propellers attached to their bottoms. By that time my breathing had returned to semi-normal (and with a conciliatory nod from one of the motorbike men) I turned my bike around and walked slowly back to the starting tent.

Just a few minutes later as I sipped on yet another violently coloured juice, still struggling to comprehend that I wouldn't be crossing the finish line with the others, the tent flap opened up and my Mauritian limped inside with a similar look of disbelief on her face. Exchanging smiles and a heartfelt high-five we sat in silent repose until another nice official adorned with several radios ushered us both outside to the designated press area where a bored looking woman from a small time German newspaper scribbled down our details. Wanting to make the most of my last few seconds of fame I informed her, proudly, that I was in fact a beekeeper – not a full time athlete to which she responded with a look of intrigue and the words, *"Jaa, das ist gut – ze bees are very important, gut job, jaaa very gut job."*

Smiling broadly I bade the German goodbye and found a prime slot in the grandstands beside my new friend to watch the end of the race.

*Yup, next time I'll do better.*

# CHAPTER FOURTEEN

In the depths of a hot, scrubby part of southern Kenya called Taita, right on the edge of Tsavo West National Park, lies one of the world's largest wildlife reserves. Covered in dense swathes of silvery green bushes, ancient lava flows and soil the color of ochre, the park is home to huge herds of dust-red elephant, lion, leopard and a number of giant crocodiles that wallow unseen in the wild, chocolaty waters of the Galana River.

Here Wagongo, a short and wiry old man possessing three crooked brown teeth, a stooped back and a pair of heavily callused hands, lives out his days in a tiny shelter made of bent sticks covered in patchy layers of mud and cow dung. Inside, there is space enough for a small fireplace and two thin mattresses, one for himself and the other for his grandson who has lived with him for several years. No one knows what happened to the parents and the boy himself is so shy that on the few occasions that I arrived at their home, he instantly melted into the background like a lost puff of wind.

Just like the estimated 1.4 billion subsistence farmers that eke out a living across the third world, Wagongo works his land every day - except Sunday, when he chooses to rest. Should the rains fall as anticipated and his seeds survive the hungry birds and later the raiding elephants he can expect to get three mixed-crop harvests a year - Inshallah. This isn't ever enough to send his grandson to school or buy them new clothes but it provides enough to keep hunger at bay in the short-term and for this he clutches the small plastic crucifix around his neck and offers up his daily thanks to the blessed Almighty.

The second central character to this story is Lucy, a tall striking English lass who has spent the majority of her adult years in Kenya occupied by what she figures is a novel solution to one of Africa's many human-wildlife conflict areas: crop raiding elephants.

Above many things, is Lucy's love for these enormous pachyderms and several years ago as a young PhD student on a day of fieldwork she witnessed a groundbreaking sequence of events that would some time later result in me, lying face down in the grass in my bee suit, pondering a luminous night sky.

One afternoon in the arid rangeland of Northern Kenya Lucy found herself parked up a short distance from a small herd of elephant gathered beneath the patchy shade of an umbrella tree sheltering from the midday heat. As the younger elephants chased their trunks around in the warm, red sand Lucy watched one of the older members of the group lumber up close to the base of the tree and luxuriously scratch its bottom on the rough, twisted bark.

Seconds later, as if released from the bowels of Satan himself came the sound of a hundred thousand wings beating above which throbbed an angry buzzing that vibrated the atmosphere.

A swarm of bees.

Before the elephants arrived the bees had been happily going about their lives zinging in and out of a nest located at the top of that very same tree, but when their world started shaking the colony was mobilized and within seconds had exploded out of the hive like a poisonous black asteroid. Picking up on the sweet smell of fury and the noise of imminent assault the elephants had scattered into the surrounding bush with astonishing speed, tails raised and ears flared in panic.

Back in the Land Rover, Lucy let out a cry of delight, confirming something she had been suspecting for a while: **that elephants are most definitely afraid of bees.**

When I met Wagongo, it was several years after Lucy had watched the elephants run in fear and several months after she had first sat down with the old man and discussed with him her ingenious plan for saving his vegetables from said crop-raiding elephants. A plan in which the elephants who were often snared, shot at or poisoned for their unfortunate penchant for staples like cabbage and maize, would also be spared. Lucy's idea, like many new things in life had taken a long time to come together but when I first met her over a coffee in a crowded café in Nairobi her idea had morphed into a solid plan.

**The concept of beehive fences to deter crop-raiding elephants** is ingeniously simple in both theory and practice and as a result she has received several prestigious awards and gained worldwide recognition.

A little star-struck, I listened closely as she described how the fences actually worked and with the aid of a short film showing a family of elephants fleeing the sound of angry buzzing, I felt something akin to excitement and hope rise inside of me.

**Yes, of course, I'd love to help out.**

\*

Beads of sweat began to appear the second the wheels touched down on the uneven surface of Gambella's solitary airstrip. Staring past the chin of my neighbor; a fierce looking black man whose forehead was covered in several rows of traditional scars, I took in the dusty yellow bush that crowded in around us from every direction.

Our short flight south from Addis Ababa had been a bumpy one, but with the occasional glimpse out the window at the breathtaking peaks and valleys of the Sheka Kafa range I was able to confirm that the next few months were going to be exceptional. A mix of fresh adventure, momentous challenge and at long last an opportunity to prove myself capable of something bigger than just me.

Gambella airport a vulgar pink building situated in the southwest corner of Ethiopia is located twenty kilometers from Gambella town, eighty kilometers from the border of South Sudan and seven hundred kilometers north of the desolate Omo River – and is a place from my memory that conjures up images of jaundiced desert horizons and nomadic tribesmen straight from the center pages of a National Geographic magazine.

Ethiopia is the world's most populous landlocked country with an estimated 96 million people of which 84% live in rural areas. The livelihood of the majority of these individuals depends on rain fed agriculture (where the average plot of land worked per family is one hectare) and from which 90% of the countries crops are produced. Ethiopia also boasts one of the largest livestock populations in the world and as a result of the poor land management practices associated with both of the above, its communities are regularly considered some of the most 'at risk', 'vulnerable' and 'food-insecure' in the world. Further statistics I pulled straight from the proposal that sat on my lap included: '40% of Ethiopia's population is considered "chronically undernourished" and over the past 30 years, the country and its neighbors (Eritrea and Somalia) have been threatened by famine at least once a decade. Severe land degradation and environmental mismanagement further squeezes the dwindling natural resource base and with shifting climate patterns set to add further instability to the equation, the road ahead lies anything but smooth.'

Slurping up the dregs of a violently-colored juice drink, whose packet was covered in the peculiar squiggles in Amharic, I quickly skimmed through the last few pages of heavy reading and as the plane bumped its way towards the terminal building I settled on the far more invigorating facts that had played a central role in my signing a lengthy contract to work in this isolated part of the continent:

*Ethiopia is Africa's largest honey producer with an estimated annual production of 45,905 tons per annum (2014) accounting for almost a quarter of* *honey produced in the continent and ranking it 10<sup>th</sup> in the world, behind Mexico, Turkey and China.*

*It is estimated that 10 million bee colonies exist in the country, the majority of which are kept in traditional beehives (made of logs and leaves) and which are strung up high in the tops of the forest trees, often several hours walk away from the beekeepers homestead…'*

The sudden 'pling' of the seatbelt sign and the resulting scrambling of sixty impatient passengers put a stop to any further reading and with a broad smile on my face I zipped my bag shut and gave the handsome Dutch boy sat across the aisle a big thumbs up.

We made it!

Whispering an excited adieu to the two Cleopatra-like airhostesses I stepped out of the plane and promptly collapsed down the stairway as a blast of hot air punched its way into each and every pore. Regaining my balance with the help of the rusting bannister I followed the crowd now milling around the entrance of the building, its half open doors offering a tantalizing view into a dark and gloomy interior.

Unfortunately, it transpired that our bags had to be collected from the edge of the runway and as the baggage tractor was still limping its way over to the plane accompanied by an alarming haze of blue-gray smoke, there was little left to do but to scuttle over to the only patch of shade offered by the tortured leaves of a heat-battered palm tree and wait.

Squinting into the glare my eyes were drawn to the tumbling scrubland fighting for the space cleared by man and his machine which, from the air, had merged together like an infinite sea of dust-yellow bush peppered with numerous circles of scorched black earth where the charcoal burners went about their work. At that point an ear splitting BANG from the tractor bought my attention back to the foreground and it was at that moment that I laid eyes on the outline of an enormous plane squatted motionless at the far end of the runway. Straining my eyes, I managed to make out the words

'WORLD FOOD PROGRAMME' stamped in immense bold lettering across its length and for the first time that day, grasped the fact that I had arrived in the kind of frontier town one normally associates with food aid people, rebel groups and weary journalists on the prowl for images of Africa's latest sob story.

Eventually our bags made it into our hands and surfing forward on a wave of perspiration, the Dutch boy (who I will now introduce to you as Sven) and I swept past a long retired x-ray machine and two snoring officials with AK47's draped nonchalantly over camouflaged shoulders. On the other side a manically waving driver, who introduced himself as Okello, loaded our things into the back of a Land Cruiser and a short while later we found ourselves hooting our way through the bustling chaos of small town Africa; an unruly arrangement of khat-smoking donkey drivers, beeping *Bajajs*, and small kiosks that sold brightly colored cloth, rip-off mobile phones, second hand shoes and a multitude of fluorescent green buckets.

Along the way we passed several fully pimped 4x4s toting satellite radio antennae, heavy duty winches and large flags that proudly fluttered their status as 'Aid Organisation.' In between the piles of wood and roadside tea shops we swerved our way past the imposing compounds of UNHCR, WFP, WHO, The Danish Refugee Council, The Swedish Refugee Council, Oxfam, Concern and the International Center for Migration: with the ongoing turbulence across the border in South Sudan, business was obviously booming.

By the time we had unloaded our bags and had taken a refreshing 'cup' shower in our eclectic new home on the far side of town, both Sven and I were ready to sample some local cuisine. Hopping aboard a tuk-tuk with our new boss in tow we clattered our way down a dimly lit street packed with men drinking beer on the pavement and straight into the center of njera, shiro and tibs heaven. Accompanied by a small glass of bright yellow Tej; a local brew made from fermented honey, we took part in several anticipatory rounds of "Cheers" until such a point arrived that none of us could take on board another morsel and we returned back home along by-then silent and deserted streets.

* * *

It always takes time to adjust to a new life in a foreign country and Ethiopia was no exception; the culture, language and customs of a fiercely proud nation; the unrelenting intensity of the sun; the questionable hygiene of most food vendors and the constant power, Internet and water outages that made navigating an alien system that much more interesting. Then there was the work itself, a new routine to follow, new expectations to manage and new goals to set within an organization hopelessly bogged with infamous red tape and government bureaucracy. It's only fair to admit that the first few weeks sped past in a dizzying blur of non-performing *farenjis* simply figuring out the what, where's and how's of life deep in the sticks.

But as I mentioned previously, it is the concept of passion, that plays the most important role in anyone's attempt to initiate progress and as the project we had signed up for was dripping in the kind of philosophies and ideals we had both long dreamt of, it didn't take long before we were able to replace that look of baffled perplexity with one of slowly evolving confidence.

The 'Gambella Eco Hub,' as the project had been termed, comprised a large strip of land along the banks of the mighty Baro River. Home to a multitude of squawking bird life, leaping Colobus monkeys, nervous bush buck and 114 native fish species including the electric catfish, the area offers visitors a glance into what this continent must have looked like long before people worked out how to make a living from chopping down trees.

Late one afternoon, after a sweaty day of traipsing about the plot identifying key land marks we had gone down to the riverbank to wash our faces free of the days accumulated sweat, dust and curdled sun cream. After a cursory glance out at the water for signs of crocodile we spotted the distant silhouette of a man paddling towards us in a traditional raft made from the hand chiseled trunk of a tree. As he approached we noticed that below his wide smile he was completely naked and that he appeared to be dragging an adult-sized body in the water behind him. Fortunately for everyone involved the man turned out to be a humble fisherman who lived his life on the opposite bank and whose rustic techniques had just hooked him one of the largest Nile Perch ever prodded by mankind.

To get to the project site itself, one travels eighteen kilometers east of Gambella center along a perfectly paved road that quickly leaves behind the slowly creeping mud and tin huts that mark the town's periphery. Beyond the numerously dotted cattle *bomas*, marked by circular thorny enclosures and multiple piles of burning manure (which keeps away the loathsome tsetse fly), one then enters a long section of barren scrubland whose soil lies black and lifeless after its annual spate of burning. Given that we had both arrived smack in the middle of the dry season - when this activity is carried out with gusto for several weeks on end, the air was filled with a sickly, yellow haze from dawn until dusk and anyone wondering around outside was subject to a thorough sprinkling of charred vegetation that floated down from the heavens and which in turn resulted in a weekly clothes washing session of equivalent intensity to a mountain stage in the Tour de France.

In the last ten kilometers before one turned off onto the hub, the land on either side of the road took on a look of more extensive abuse; a result of the work of several 'land investors' well-renowned for their dodgy development deals that inevitably involved the proposing of hefty agriculture projects to high-flying ignoramuses in government. Once they had received the official nod (facilitated by the appropriate method of 'palm greasing'), various bulldozers would be commissioned to sweep across the land felling every tree, bush and grass-roofed *tukel* in sight. Those objects not able to move out the way fast enough would then be dragged into one enormous pile and set alight and in the first few weeks alone, our daily commute forced us to watch as acre-by-acre, huge swathes of Tamarind, Acacia and Amarula trees (plus all the critters and granules of soil nutrient in situ) were swallowed in a weeklong wall of ravenous flames. A few days later, once the coals had stopped glowing and piles of knee-deep ash lined the roadside, it was alleged that an unknown man wearing dark sunglasses and clutching a soon-to-be-filled briefcase would stop by and click a few photos, confirming that 'progress' was indeed taking place and which would in turn result in the release of substantial loans into foreign bank accounts… and indeed, several months down the line we are yet to bear witness to any further activity.

Awesome, humanity.

*

But for the first time both Sven and I had a way of defending ourselves against the crushing bombardment of negativity associated with our species treatment of the earth, for we now had a plan and a long dreamt of opportunity to process the amassed collection of ideas that fell beneath the banner of possible solutions. Along with the comforting knowledge of a regular salary and much enthusiastic communication with our Dutch funders, it was impossible not to feel honored that the universe had – at last - considered us ready for such a task.

At its core the 'hub' was a project that fully embraced the term 'earth stewardship' by taking a designated area of land in a remote, much abused part of Sub-Saharan part of Africa and turning it into a showcase of ecological integrity; a multifunctional center whose overriding goal was to present attractive alternatives to prevailing farming methods that were both ecologically sound and economically productive.

Thanks to the bees this concept was not entirely new, but nor was it for a global movement of people who had, over the decades defined and implemented the guiding principles of permaculture; a method of adapting and living with the earth that has begun to gather momentum in city streets and farmers' fields all over the world and whose basic definitions stands as:

PERMACULTURE:
- a holistic design system for creating sustainable human settlements and food production systems.
- a movement concerned with sustainable, environmentally sound land use and the building of stable communities, through the harmonious interrelationship of humans, plants, animals and the Earth.

Since its humble beginnings on a small farm in SW Australia in the late 70's, the term "permaculture" stemming from the simple combination of the worlds 'permanent' and 'culture' sets a new command for an ancient set of guiding principles and associated techniques that had blown apart my brain with an explosion of fresh ideas, ageless wisdom and basic common sense.

With Sven as my guide (for he had been a devotee for some time and was now an avid collector of exotic fruit and vegetable seeds) I was at last able to ground many of my own rustic beliefs about the earth, its inhabitants and our collective responsibility to safeguard its future.

Thus, once we had both secured our Ethiopian driving licenses (a process that took several weeks and multiple interviews with suspicious officials convinced of espionage), we were able to flee the cramped Gambella office and work began immediately. Hand picking our work force; a slightly rag tag group of young teenagers from the neighboring Jawe community who sported clip on earrings, sagging boxer shorts and a Jay-Z inspired 'slouch', we set about one particular area on the edge of a rocky outcrop that boasted immense views of the river and its surrounds; digging holes, testing the soil and mapping out the site to include a number of buildings and demonstration features such as herb spirals, food forests and banana circles.

Initially progress was slow, but on site, far from the chaotic bustle of town and its scheming business men (whose prices soared the second they laid eyes on the fluorescent glow of our foreign skins), a large Anuak-style *tukel* began to take shape along with a set of open air showers, several composting toilets and a mud walled storeroom that one day I envisaged transforming into a beautiful honey room and community training center. It didn't take long for Sven too, to work his magic amongst the barren patches of rain-starved earth and within a matter of weeks there was an outburst of fragile green shoots, the odd waft of basil and the refreshing tang of rocket emanating from a heat-soggied sandwich.

With daily temperatures forcing the mercury to hover up around 45 degrees the working days were long and arduous and we would return home each evening exhausted, filthy and blistered. But every now and again, in between the clang of spade-meets-rock, the incessant hammering of wood and the encouraging "YALLA YALLA" shouts from our amateur work force, one could glance up and take in the array of sights and sounds of the natural world that buzzed, chirruped and rustled all about us. Once we had established a permanent water supply on site this effect appeared to multiply as each plant, bug, insect and animal swooped in to take advantage and with a lengthy break over the hottest hours at noon there was little else to do but to stretch out on bamboo mats and take it all in.

This uncluttered way of life, I swiftly discovered, had long been perfected by the local people around us who rose early to fish, hunt and gather in the coolest hours of the day and who would often arrive on site ahead of a long day of digging with a bag of sweet yellow berries tucked beneath their arm or a battered USAID tin filled to the brim with wild honeycomb – and into which we would be invited to plunge our fists just like Winnie the Pooh. Exchanging the universal thumbs up sign for 'wicked' we slowly learnt to relax in one another's presence and with the constant translation and exchanging of techniques on everything from harvesting honey to mixing concrete we soon identified some particularly talented individuals.

They would become our 'champions.'

And as the weeks gathered pace, merging into one continual morphing of big ideas I began to note something else too; a conscious relaxing of my grip on the incessant search for answers - the "where am I goings?" and "what do I wants?" that had hounded me for so long suddenly became redundant as I found the time to absorb the full scope of what had so miraculously landed up in my lap – a chance to prove myself - just as I had asked for all those months ago.

And with that came a novel sense of contentment – of quiet relief.

Something out there heard my voice – It hears us all.

* * *

Mangos, sugar cane, papayas, sweet potatoes, moringa, strawberries, chilies, guava, baby tomatoes, and lettuce; a never-ending buffet of creepers, tubers, roots and leaves that we purchased as young seedlings, planted from seed or 'liberated' from roadside verges and abandoned homesteads.

**Food, glorious food.**

But it was about so much more than that, for as each day passed I felt myself growing from the inside. Suddenly I found the incentive to look deeper into the world at my feet and above my head, a world where the honeybee existed as one tiny part of a much bigger web of connection and balance.

**Just as I was too.**

Our mandate at the slowly evolving hub went beyond just pressing seeds gently into the soil and eating the end result. Rather it went into designing an entire living system together with the local people who had witnessed with their own eyes the slow degradation of their land over the years and whose mindset we eventually hoped to transform with patient demonstration and practical philosophy.

But life without challenge - even on this miniature scale, would soon become boring and as my deadline for this book looms we are just four months into this tiny project of enormous ambition and the battle to succeed, prove, change has barely even begun. For over and above the natural obstacles that are sent forth to test our determination on a daily basis; the stubborn soil, the obstinate aphids and the continual absence of rainclouds, are the human challenges too, which, more often then not, are more disruptive to simple hopes and untested dreams then anything else.

**Can we do it? – Is there any point? – What if we don't succeed?**
But this time around I am fully prepared – for this is my time, OUR TIME – and in that very same instance comes a small but determined voice from somewhere deep within:

**Breathe. Calm. Focus. Begin.**

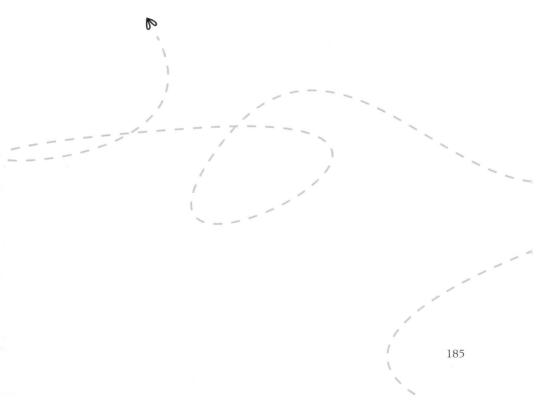

# Pictures from Jess

Spot the difference? 5+ members of the Oloika women's group who ventured out to watch the white lady climb trees in 40°C with a smoker...

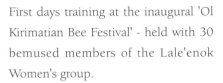

First days training at the inaugural 'Ol Kirimatian Bee Festival' - held with 30 bemused members of the Lale'enok Women's group.

Looking naively merry before a warm-up race ahead of the commonwealth games. 4 hours later and the cheer was somehow less evident!

# ABOUT THE AUTHOR

Jess de Boer was born and raised in Nairobi, Kenya to a privileged Dutch family in the region. From a young age, inspired by her surroundings, Jess de Boer had a strong desire to bring about appositive change in the world. On graduating from University, Jess worked in a variety of industries, ranging from being a private chef in the Swiss Alps to maggot farming in Thailand, while travelling the world and documenting her many attempts to make a positive change.

In 2014 Jess won The Africa Book Club Short Reads competition with her story The Honey Man. That same year, she represented Kenya in the Women's Triathlon at the Commonwealth Games in Glasgow. She now works as a beekeeper in Kenya and is still on a mission to one day save the world. The Elephant and the Bee is her first book.

# BIBLIOGRAPGHY

Callum Roberts, *Oceans of Life. How are seas are changing. London:* Penguin Books, 2012.

Generation Jobless, The Economist, [online], available from: http://www.economist.com/news/international/21576657-around-world-almost-300m-15-24-year-olds-are-not-working-what-has-caused

M.F. Johannsmeier, *Beekeeping in South Africa 3rd Edition.* Pretoria; Plant Protection Research Institute (Planr Protection Handbook No. 14) (2001)

Krulwich Robert, NPR, *How Important Is A Bee?* December 06, 2013 http://www.npr.org/sections/krulwich/2013/12/04/248795791/how-important-is-a-bee

Sandra Yin and Mary Kent, Population Reference Bureau, *Kenya: The Demographics of a Country in Turmoil.* http://www.prb.org/Publications/Articles/2008/kenya.aspx

*Noah W. Wawire & Gerald R. S. Ochiel, Review of the impact of water hyacinth on Lake Victoria: The case of Winam Gulf, Kenya.*

Ambrose Evans-Pritchard, Einstein was right - Honey bee collapse threatens global food security, The Telegraph, [online], available from: *http://www.telegraph.co.uk/finance/comment/ambroseevans_pritchard/8306970/ Einstein-was-right-honey-bee-collapse-threatens-global-food-security.html*

Kluser S., Neumann P., Chauzat M., Pettis J.S., Peduzzi P., Witt R., Fernandez N. & Theuri M., Global honey bee colony disorders and other threats to insect pollinators, 2010.

Moorsom Pierre, Permaculture in Malawi, The Guardian, [online], available from: *http://www.theguardian.com/global-development-professionals-network/ 2015/apr/20/permaculture-malawi-food-forests-prevent-floods-hunger?CMP= share_btn_link*

Hove H., Echeverria D., Parry J. and International Institute for Sustainable Development, Review of Current and Planned Adaptation Action: East Africa. Adaptation Partnership, [online], available from: http://www.cakex. org/virtual-library/review-current-and-planned-adaptation-action-east-africa

ILO (2013), Global Employment Trends For Youth 2013, A Generation at Risk.

Bass S, et al. (2013), "Making Growth Green and Inclusive: The case of Ethipia", OECD Green Growth Papers [online], No. 2013/07, OECD Publishing, Paris, http://www.oecd-ilibrary.org/environment/making-growth-green-and-inclusive_5k46dbzhrkhl-en

USGS (2012), A Climate Trend Analysis of Ethiopia. Famine Early Warning Systems Network – Informing Climate Change Adaptation Series.

# USEFUL WEBSITES

*www.agriprotein.com*
*www.elephantsandbees.org*
*www.ecopost.co.ke*
*www.realipm.com*
*www.iimsam.org*

# ACKNOWLEDGEMENTS

There are many, many people out there who deserve a massive and profound thank you, so here it is.

*To my agent Jayapriya Vasudevan of Jacaranda Press for your brilliant laugh and unwavering confidence in this book. You are a truly glorious soul.*

*To the utterly magnificent team at Jacaranda Books: Valerie, Jazzmine, Cynthia and Janneke, I don't think I will ever forget the first time we met and let this stand as an eternal IOU for multiple jars of homegrown honey. Thanks to Joe for the epic designs and to Rukhsana who set me on my way.*

*Big up to Travis and Riva Reynolds for reading the manuscript through and all of my glorious friends who replied to my random demands for advice and assistance.*

*To my family, no words can express what you are to me. An appalling tribute really but I don't even know where to start.*

*To Sven Verwiel for putting up with me in that place and at that time in the morning for so long. You did good, kid, and there are so many unwritten chapters ahead.*

*And lastly, to all the people out there who are simply trying to make this world a more marvelous place. Keep doing what you are doing. We have a bumpy ride ahead of us but I believe with every cell in my brain that your actions – no matter how small or insignificant they may feel at the time, absolutely matter.*